from
the
black
hole
to
the
infinite
universe

goldsmith
levy

from
the black hole
to the
infinite universe

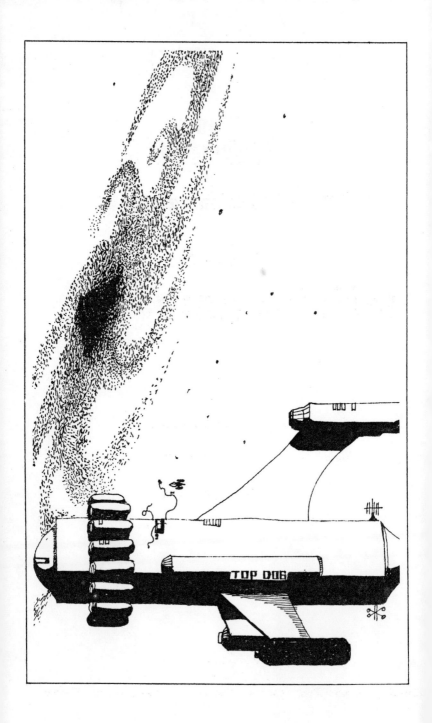

from the black hole to the infinite universe

donald goldsmith

State University of New York at Stonybrook

donald levy

University of California at Berkeley

Holden-Day, Inc.

San Francisco

Düsseldorf Johannesburg London Panama Singapore Sydney Toronto

Editor: Eva Marie Strock
Designer: Gillian Johnson
Illustrators: Renaldo Ratto, III (chapter frontispieces);
        Donald Goldsmith, Francis Quock,
        and Lawrence Anderson
Cover designer: Gillian Johnson
Production: Charles A. Goehring
Composition: Applied Typographic Systems, Inc.
        (Century Schoolbook)
Printing and binding: BookCrafters, Inc.

FROM THE BLACK HOLE
TO THE INFINITE UNIVERSE

Copyright © 1974 by Holden-Day, Inc.
500 Sansome Street, San Francisco, California 94111

Library of Congress Catalog Card Number: 73-86412
ISBN: 0-8162-3323-3

1234567890   BC   807987654

Printed in the United States of America

contents

from
the black hole
to the
infinite universe

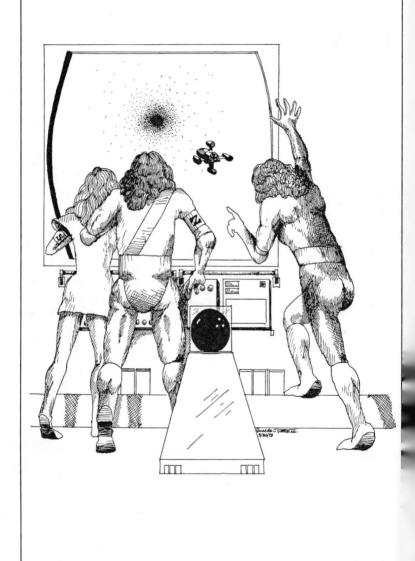

1
black
holes

Cyril Zaki's huge hands trembled slightly as he
adjusted the gyrocontrols of his minispacecruiser. His
calm features were disturbed by a tic at the corner of his
right eye. He and his ship the Top Dog were about to
penetrate a totally black region of space. On his twenty-
one-centimeter maps, dark green lines marked this black
hole as forbidden ("avoid unless unavoidable"). During
the last accounting period, four space liners and five
trading vessels had disappeared in this dark area,
never to be seen again. Not a single distress message
had been heard. Radar waves beamed into the region
never bounced back but were swallowed up like the
spaceships. Despite his years as a space insurance
investigator, Zaki was baffled by the blackness in front
of him.

With one hand he lit a melloroon, pure coroneel
extract from the bahava plants on Heiles IV. Smoke too
many and you get a bad heart, Cyril thought, but that
was better than being a zoroo like Boyer, swallowing
pills all day and imagining three lives at once. Boyer,
Zaki reflected, was also one of a growing number of
space crewmen who had let their premiums lapse. Cyril
realized that this was not the time to worry about
nonconformists. He leaned forward and pressed the
orange button to activate the ship's intercom system.

4 *Years as a top insurance investigator for Guaranteea Trust paid off as he spoke into the intercom with a tone of cold command:*

"Officers Zots, Card, Dalby, and Boyer to the bridge on the double."

Zaki took a hit off his melloroon while waiting for his officers to appear. As he glanced around the control room, he felt a certain pride as he noted each priceless and well-insured piece of equipment occupying its proper place. In the center of the ship's bridge, resting on a silver pedestal, stood a shiny black sphere with three holes in it, the use of which even Cyril couldn't fathom. Like all the other irreplaceable equipment on board, it too was fully insured. Zaki raised a slightly shaking hand to his lips and took another deep drag on his melloroon. He forced his mind back to the problem at hand, which he wanted to review mentally before his crew arrived on the bridge. The arrival of his officers interrupted him in midthought. Zots, Card, and Dalby looked worried, but Boyer sauntered in with a cool gaze, humming a popular space melody: "You're a bad risk, baby, ... But I'd put a big premium on you." His eyes had the characteristic zoroo glaze. Cyril wondered what could be wrong with the times, if even dedicated space officers were letting their premiums lapse. With some difficulty, he wrenched his thoughts from Boyer and the troubled times and spoke.

"Dalby, focus the external viewers on the black hole region, and you, Zots, prepare an unmanned drone for launch."

The large viewscreen came to life, only to reveal a jet-black area: no starlight, no suns, no debris, nothing. Second Officer Zots asked:

"Shall I launch the probe, Captain?"

Cyril lit another melloroon.

"Yes," he said. "And you, Dalby, get ready to follow it with the viewers. I want to see just what we're up against."

Zaki recalled the time that he had avoided landing on a planet made of mush by cleverly deploying a probe like this one.

"Drone away, Captain," said Zots.

The small silver dart left the Top Dog's underbelly and headed for the black hole. A tense group followed its progress on the viewscreen. As the minutes sped by, the drone accelerated toward the black hole region, seemingly unharmed. Suddenly a collective gasp swept the bridge. Turning red for an instant, the drone suddenly vanished. The crew was stunned into silence, except for Boyer, who remarked "Out of sight."

"Great cancellation," murmured Lisa Dalby, "What's out there? Some kind of mammoth garbage disposal?"

Zots was visibly shaken. "You don't suppose the drone entered another space-time manifold, do you?" he asked.

Zaki didn't know, but he did know that he was the captain. He also knew that Lisa Dalby had the finest pair of legs he'd seen in space. What kind of beings could have made such a cosmic sump? Who were its intended victims? Visions of huge worms with human arms and smiling faces leapt into Zaki's brain, unwelcome remnants of his childhood on Herpes. Through his haze of conflicting thoughts, Cyril could sense the panic mounting among his crew. He wanted to calm them, but the words wouldn't come. An eerie silence hung in the small, cluttered cabin for several minutes, until Julie Card reported:

"No radar signals bounce back, Cyril, nothing. The drone's completely gone. Not a trace."

The only person on the bridge seemingly unaffected by all this was Sepp Boyer, who continued his quiet humming, oblivious to the others' apprehensions. But in a movement rapid for a zoroo, he turned his head from the alpha control panel and said:

"Captain, my E meter shows we're being accelerated toward the black hole at three kilometers per second per second."

6   *Now even Boyer looked worried. He turned back to hunch over the E meter.*

Zaki lit a third melloroon, then realized that he hadn't finished his second. With a deft and rapid movement of his thumb and forefinger he crushed his third melloroon and swept up the second from its niche on the alpha panel. Per second per second per second percolated through his mind. He had to act before this thing acted on him.

"Julie, engage the thrusters. Ten units of power in reverse," Cyril ordered.

Boyer reported the effect quickly. "We're not accelerating toward the hole now, but we're not moving away either."

Zaki instantly told Card, "Julie, give us all twelve units." The limit, and everyone knew it.

Boyer became deeply involved with his E meter. Finally he spoke.

"We're accelerating away from the center of the black hole now. We should be clear in five minutes."

The crew on the bridge was visibly relieved.

Good risk, Cyril thought, another five minutes at twelve units and the thrusters will surely need repairs. He pulled himself together and asked:

"Julie, what's the nearest planet? We'll have to land for refitting the tubes. We've verified that the hole exists, that's all we can do now. We'll write a number two memo when we land."

Zots whispered to Lisa Dalby, "That's our Captain, all right, Mister essence of exploration. Imagine, a number two memo."

Lisa ignored Bernie Zots. She realized that Cyril wasn't perfect, but he was gentle for a big man, and there were times when being gentle was more important than being smart. A smile crossed her lips.

Julie Card spent a few minutes with the twenty-one-centimeter maps and answered. "The nearest is Sidney, at the outer edges of the dark region. It's an old pleasure planet but off limits to insurance personnel. Parts of it are actually uninsured."

Cyril inhaled on his melloroon, coughed several times, and recalled that the name Sidney had come up in conversations with his boss, Zenith Borg. He'd always wondered how it could be uninsured. Well, there was no help for it.

"Lisa," he muttered, "Put the Dog on course for Sidney. I'm taking full responsibility for the decision."

8 Could a black hole really exist? What kinds of forces could produce such an object? A totally black hole like the one Captain Zaki encountered must attract absolutely everything. Are there any forces that attract everything? There is one such force that is so much a part of our lives that we tend to forget it: *gravitation.* The most important thing about gravity is that gravitational forces *always attract* one piece of matter to another. The matter can be of any size or form; it may consist of individual atoms or more complicated combinations of atoms such as molecules, plastics, people, or planets.

The force of gravity that acts between any two chunks of matter grows stronger in proportion to the amount of material (we call this the mass) that forms the chunks, and it also grows stronger if the centers of the two chunks get closer to one another (Fig. 1). The moon, which has less mass than the earth, exerts less attractive gravitational force on other objects than the earth does, which is why astronauts "weigh" less on the moon than on earth: They feel a weaker force of gravity pulling them down.

How does gravity act on forms of electromagnetic energy like light and radio waves and radar beams? A black hole would attract them. In 1917, Albert Einstein realized that gravity would also attract the particles (*photons*) that form light and radio waves and radar. The gravitational force from a massive body will attract photons just as it attracts ordinary particles. When a

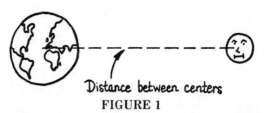

Distance between centers
FIGURE 1
Two typical massive objects are attracted to each other by the
force of gravity, which acts along the line between their centers.

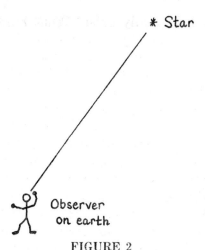

* Star

Observer
on earth

FIGURE 2

beam of light from a distant star passes close to a star like our sun, the light beam is bent toward the sun by the sun's attractive force of gravity (Figs. 2 and 3).

We can observe this gravitational bending of light rays in the following way. First, we determine the locations of some distant stars at a time when the sun is not in front of the area of the sky where the stars are (Figs. 2 and 3). Later, at a time when the earth's motion around the sun has caused the sun to lie between the earth and these stars, we can remeasure the position of the stars[1] to determine the apparent positions of the stars in the sky. Such measurements were made during an eclipse in Africa in 1919 and provided an important confirmation of Einstein's relativity theory because they showed that light rays *are* bent by the sun's force of gravity, just as Einstein had predicted.

We have seen how light rays passing near the sun are bent toward it by the gravitational attraction of the sun. The same gravitational forces also act on any light ray

---

[1]Because the sun is so bright these measurements can be made only during a total solar eclipse.

10

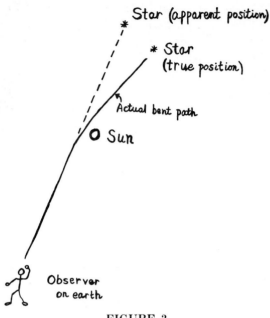

FIGURE 3

that tries to escape from the sun's surface (Fig. 4). Such light rays must overcome the sun's gravitational attraction before they can reach us here on earth. This situation is analogous to the astronauts' trip from the earth to the moon because their rocket must overcome the earth's force of gravity before it can leave the earth and travel onward toward the moon. Any comet or asteroid that passes near the sun will be pulled toward it by gravitational forces in the same way that light rays are because as mentioned, gravity attracts everything: comets, light rays, rocks, radar, and us.

What would happen if the force of gravity at the surface of the sun (or of any star) were to grow billions of times stronger? Light rays passing near the sun would be bent right into it in ever-tighter spirals, and they would be totally absorbed. Comets, asteroids, or space ships that came too close to the sun would be pulled into the sun too. Also, any light rays trying to escape from the sun's surface would be pulled back into the sun by the increased forces of gravitation (Figs. 4 and 5). At this point, radar, radio, or light beams directed at the sun would not bounce back since they, like everything else, would be caught and held by the force of gravity. Thus, *any* kind of matter or radiation that came too near the sun's surface or tried to leave the surface would be caught up and bound to the sun. With no light able to leave its surface, the sun would appear to be black; in fact, it would then be a black hole.

We still need to understand how the gravitational force at the surface of the sun or any other star might grow strong enough to create a black hole. Recall that gravity is an attractive force that acts between any two chunks of matter (Fig. 1). The gravitational force is more powerful if the chunks have more material (mass) and if their centers are closer to one another.

To illustrate these principles and to understand how a black hole behaves, imagine an indestructible person

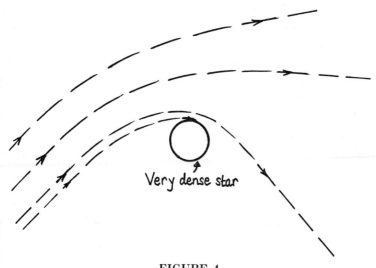

Very dense star

FIGURE 4
Light rays passing close to a very dense star will be bent right into it.

Very dense star

FIGURE 5

Light that tries to escape from the surface of a star must overcome
the force of gravity. If the star is very dense (highly contracted),
light will not escape from its surface at all.

who can stand on the surface of the sun. We can consider
the person as one chunk of matter and the sun as another.
The distance between the sun's center and the person's
center (the person's stomach) is the radius of the sun, or
about 400,000 miles. This situation is drawn in Fig. 6a.

Now suppose that the sun were to collapse, so that its
radius became much smaller than it really is (Fig. 6b).
If this collapse were to occur without the sun losing any
of its mass, the force of gravity on the person at the sun's
surface would increase because the distance between the
person's stomach and the center of the sun would decrease.
How small a radius would the sun need to have before
the gravitational force at its surface became so strong
that it would make the sun a black hole? The answer is
about 2 miles. A person on the surface of a star like the
sun that had contracted to a radius of 2 miles would feel
a force of gravity about a trillion times stronger than the
force of gravity she would feel on the earth's surface now.
This immense gravitational force would be strong enough
to keep any light rays, radio waves, or anything else
from leaving the collapsed star's surface, and anything
that came within 3 miles of the star's center (1 mile of its
surface) would be pulled into this black hole, never to
emerge.

Have any stars undergone contraction under the force
of gravity to the point where they have become black

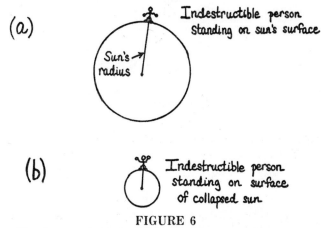

(a)

Indestructible person
standing on sun's surface

Sun's radius

(b)

Indestructible person
standing on surface
of collapsed sun

FIGURE 6
(a) The distance between the sun's center and the center of an
indestructible person on the sun's surface is the radius of the sun.
(b) If the sun were to collapse, the distance from the sun's center
to the center of the indestructible person would decrease.

holes?[1] The question is difficult to answer because the
nature of black holes makes them invisible. The gravitational force from a black hole never stops; it continues to
act on objects that are both too far away from the black
hole to be caught and held to its surface and close enough
to be captured by the black hole. For example, if our sun
became a black hole, the earth would continue to orbit
around it (Fig. 7) because the earth would still feel the
sun's attractive gravitational force. In its present orbit
the earth is too far from the sun to be swallowed up even
if the sun became a black hole and light no longer left
its surface.

If a black hole were located in the vicinity of a visible
star, but not close enough to capture the star, the black
hole could cause the star to orbit around it because of its
gravitational force. (Actually, the star and the black hole

---

[1]The critical radius at which a star becomes a black hole varies in direct
proportion to the amount of material (mass) in the star. For the sun, and for
any other object with the same amount of mass, the critical radius is 2 miles.
For a star with twice the sun's mass, the critical radius is 4 miles, and so forth.

14

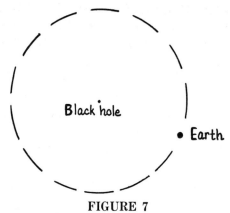

FIGURE 7
If the sun became a black hole, the earth
would continue to orbit the black hole.

would orbit around their common center of gravity,
which would lie halfway between the center of the star
and the center of the black hole if the star and the black
hole had the same amount of mass.) The motion of the
visible star around the black hole might show up as a
wobble in the position of the visible star, and we could
detect this wobble even though we could not see the
black hole that was causing it (Fig. 8). There are a few
preliminary indications that some stars, such as one
called Cygnus X-1, may have a wobble in their motion
through space which is produced by a black hole in their
immediate vicinity. If Cygnus X-1 and a black hole were
indeed in orbit around each other, the black hole would

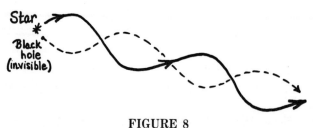

FIGURE 8
A black hole in the vicinity of a star would cause a
wobble motion in its drift through space.

tend to pull matter out of Cygnus X-1. This matter could **15** be seen before it vanishes forever into the black hole and could be producing the x-rays that are being emitted in the vicinity of Cygnus X-1. The present evidence is not decisive, and black holes may yet prove to exist only in theory. On the other hand, it is entirely possible that black holes have consumed *most* of the matter in the universe, which would mean that the stars and galaxies we see now actually form only a small fraction of the universe's total mass. This would change the predicted future of the universe in several ways; for example, the black holes might contain enough mass to reverse the present expansion of the universe and eventually produce a contraction of the universe under the force of gravity.

★ ★ ★ ★ ★ ★ ★ ★ ★ ★

The theory of gravitation has an elegant simplicity. We can understand most of it by using the simple algebra of Newton's law of gravitation, which gives us most of our knowledge about the force of gravity between any two objects.[1] The law states that the gravitational force of attraction between any two bodies acts along the line between their centers. In addition, this force is always proportional to the product of the masses of the two bodies, and it is inversely proportional to the *square* of the distance between their centers (Fig. 1). Thus, using the sign $\propto$ to represent proportionality, we may write

$$\text{Force} \propto \frac{(\text{mass of one body}) \times (\text{mass of other body})}{(\text{distance between centers})^2}$$

If we choose the units in which we measure the masses and distances, we can change the sign of proportionality

---

[1]Newton's law holds only for the force between two bodies that have mass. A different law gives the force of gravity that acts on light and radio waves because the particles (photons) that make up these waves have no mass.

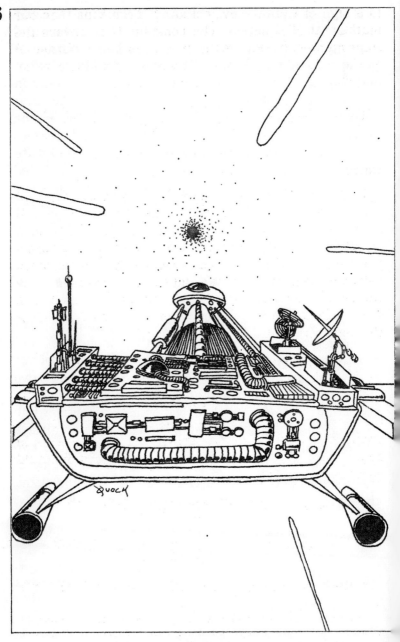

FIGURE 9

to a sign of equality by including a constant G in our
mathematical equation. The constant G is always the
same number. It changes the units of (mass)^2/(distance)^2
on the right-hand side of the equation into units of force,
and Newton's law of gravitation becomes[2]

$$\text{Force} = \frac{G \times (\text{mass of one body}) \times (\text{mass of other body})}{(\text{distance between centers})^2}$$

To understand Newton's law of gravitation more fully
we shall discuss four specific examples:

1 A person's weight equals the force of gravity that
acts on her. On the earth's surface a person's weight will
be just the force of the earth's gravity upon the person.
To calculate the force of the earth's gravity acting upon a
person at its surface, we use Newton's law. The distance
between the person's center and the earth's center is the
radius of the earth, about 4,000 miles (Fig. 10a). The
gravitational force of the earth on the person, which is
the person's weight, is

$$\text{Weight} = \frac{G \times (\text{earth's mass}) \times (\text{person's mass})}{(4,000 \text{ miles})^2}$$

If the *mass* of the earth were twice what it is now, but
the eartn's radius and the person's mass were to remain
unchanged, then the force of gravity on the person would
double because of the greater mass of the earth. We could
write this result as

$$\text{Force} = \frac{G \times (2 \times \text{earth's mass}) \times (\text{person's mass})}{(4,000 \text{ miles})^2}$$

$$= 2 \times \frac{G \times (\text{earth's mass}) \times (\text{person's mass})}{(4,000 \text{ miles})^2}$$

2 Now suppose that the earth's mass and the person's

---

[2]If we measure masses in grams and distances in centimeters, the value of the
constant G is always $G=6.67 \times 10^{-8}$ centimeter^3/(gram-second^2) [cm^3/(gm-sec^2)].
In general, though, we shall deal with algebraic relations in such a way that
we do not require a knowledge of the specific units involved.

18   mass were to remain the same but the earth's radius were to double, to 8,000 miles (see Fig. 10b). The person's weight would then be reduced to one-fourth its original value because the distance between the earth's center and the person's center would increase. The algebra of this situation is

$$\text{Force} = \frac{G \times (\text{earth's mass}) \times (\text{person's mass})}{(8{,}000 \text{ miles})^2}$$

$$= \frac{G \times (\text{earth's mass}) \times (\text{person's mass})}{(2 \times 4{,}000 \text{ miles})^2}$$

$$= \frac{1}{4} \times \frac{G \times (\text{earth's mass}) \times (\text{person's mass})}{(4{,}000 \text{ miles})^2}$$

Both these examples illustrate that Newton's law of gravitation tells us not only that the force of gravity increases with mass and decreases as the separation of centers increases, but it also tells us the amount of the increase or decrease.

3   We can also use Newton's law to show that the force of gravity acting on an indestructible person standing on the sun's surface would be 30 times greater than the force of gravity acting on the same person standing on the earth. To do this, we need to know that the sun's radius (400,000 miles) is 100 times the earth's radius (4,000 miles) and that the sun's mass is 300,000 times the earth's mass. We turn now to Newton's law.

The gravitational force on a person at the sun's surface is

$$\text{Force} = \frac{G \times (\text{sun's mass}) \times (\text{person's mass})}{(\text{sun's radius})^2}$$

Using the values for the sun's mass and radius in terms of the earth's mass and radius, we find that this equation is

$$\text{Force} = \frac{G \times (300{,}000 \times \text{earth's mass}) \times (\text{person's mass})}{(100 \times \text{earth's radius})^2}$$

$$= \frac{300{,}000}{10{,}000} \times \frac{G \times (\text{earth's mass}) \times (\text{person's mass})}{(\text{earth's radius})^2}$$

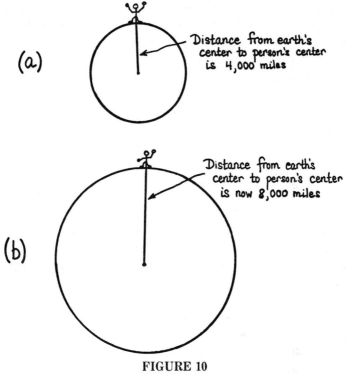

(a)

(b)

FIGURE 10
(a) The distance from the earth's center to the earth's
surface is the earth's radius, about 4,000 miles.
(b) If the earth's radius were to become twice as large as
it is now, the distance from the earth's center to the center of
a person standing on the earth's surface would be twice as great.

Because $\frac{300,000}{10,000}$ equals 30, we see that the force of
gravity on the sun's surface is 30 times the force of
gravity on the earth's surface. Thus a man who weighs
150 pounds (lb) on earth would weigh 4,500 lb on the
sun's surface.

4   Why doesn't the sun's gravitational force affect the
weight of a person standing on the earth? It does, but
only by a tiny amount. The distance of the earth from
the sun is 93 million miles, or about 23,000 times the
earth's radius (Fig. 11). Therefore, the force of gravity
from the sun on a person on the earth's surface is

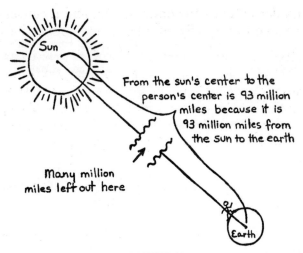

FIGURE 11

The distance from the sun's center to the earth is about 93 million miles. This distance is about 23,000 times larger than the radius of the earth. Therefore, a person on the earth's surface is 23,000 times nearer the center of the earth than the center of the sun.

$$\text{Force} = \frac{G \times (\text{sun's mass}) \times (\text{person's mass})}{(\text{earth-sun distance})^2}$$

$$= \frac{G \times (300{,}000 \times \text{earth's mass}) \times (\text{person's mass})}{(23{,}000 \times \text{earth's radius})^2}$$

$$= \frac{300{,}000}{(23{,}000)^2} \times \frac{G \times (\text{earth's mass}) \times (\text{person's mass})}{(\text{earth's radius})^2}$$

$$= \frac{1}{15{,}000} \times \frac{G \times (\text{earth's mass}) \times (\text{person's mass})}{(\text{earth's radius})^2}$$

$$= \frac{1}{15{,}000} \times \text{Force of } \textit{earth's} \text{ gravity}$$

The force of gravity of the sun upon a person on the earth is less than one ten-thousandth of the gravitational force that the earth exerts upon the person. For this reason we can ignore the force of gravity from the sun in comparison with the earth's force of gravity when we are located on the earth's surface. Of course, the sun's force

of gravity does attract the entire earth, and it causes the
earth to move in a nearly circular orbit around the sun.

Now that we have seen how Newton's law of gravitation works we can consider how the sun's gravitational force would change if the sun were to shrink to the size of a black hole. If the sun shrank to a radius of 2 miles without losing any of its mass, an indestructible person on the sun's surface would feel a gravitational force given by

$$\text{Force after shrinking} = \frac{G \times (\text{sun's mass}) \times (\text{person's mass})}{(2 \text{ miles})^2}$$

We can now compare this with the force of gravity on the sun's surface:

$$\text{Force before shrinking} = \frac{G \times (\text{sun's mass}) \times (\text{person's mass})}{(400{,}000 \text{ miles})^2}$$

To find the increase in the gravitational force on the person at the sun's surface after the contraction of the sun, we take the ratio of the two forces:

$$\frac{\text{Force after shrinking}}{\text{Force before shrinking}} = \frac{G \times (\text{sun's mass})}{G \times (\text{sun's mass})}$$

$$\frac{\times (\text{person's mass})/(2 \text{ miles})^2}{\times (\text{person's mass})/(400{,}000 \text{ miles})^2} = \frac{1/(2 \text{ miles})^2}{1/(400{,}000 \text{ miles})^2}$$

$$= \frac{(400{,}000)^2}{(2)^2} = 40{,}000{,}000{,}000$$

The gravitational force at the sun's surface would increase 40 billion times; this would produce such an enormous force that the sun would become a black hole. Note that the gravitational force at the *surface* of the sun is what increases as the sun shrinks. A person at a constant distance from the sun (for example, on the earth, which moves in an almost circular orbit around the sun) would not experience any great change in the gravitational force if the sun were to collapse without losing any mass because the force depends only on the person's mass, the sun's mass, and the distance between their centers, and none of these quantities change as the sun shrinks.

SUMMARY

Gravitational forces always *attract*. The gravitational force between any two objects increases if the objects have more mass (mass corresponds to our intuitive notion of weight), and the force decreases if the centers of the two objects are farther away from one another. Gravity also acts on light rays. If a star collapsed to produce an object of extremely high density, this object could have a gravitational force so strong that *nothing*, not even light, could escape from its surface. Such collapsed objects, called black holes, may indeed exist. They could make their presence felt only through the gravitational force they continue to produce.

QUESTIONS

1 Does the earth's force of gravity attract hair? Plastic? Spaceships?
2 Light that is produced on the surface of stars like our sun leaves the stars and travels through space to the earth, where it enters our eyes and allows us to see the stars. Why are black holes invisible?
3 If the mass of the earth were to triple, and the earth's radius remained the same, would we weigh more or less? By how much? (*Ans.*: three times more)
4 Does a person on Mount Everest weigh more than a person in Death Valley, or does she weigh less?
5 If the sun's radius were to shrink until the sun became a black hole, how much would the sun's force of gravity on the earth be in comparison to what the sun's force of gravity is now?

6 If a man feels a force of gravity of 180 lb and his wife feels a force of gravity of 90 lb, who is the more massive? By how much?

7 The mass of the planet Mars is one-tenth the mass of the earth, and Mars' radius is one-half the radius of the earth, approximately. How much would a 120-lb woman (on earth) weigh on Mars? (*Ans.*: 48 lb)

8 The radius of the moon is about 1,000 miles, and a man on the moon feels a force of gravity that is one-sixth the force of gravity he feels on the earth's surface. The earth's radius is about 4,000 miles. About what fraction of the earth's mass is the moon's mass? (*Ans.*: $\frac{1}{96}$)

9 Suppose that women landed in a spaceship on the planet Jupiter, which has a mass 300 times the earth's mass and a radius 10 times the earth's radius. The force of their rocket motors must balance the force of the planet's gravity. How much more force would the women need to balance Jupiter's gravitational force at its surface than they need to balance the earth's gravitational force at the earth's surface? (*Ans.*: three times more)

2
atoms
to atoms

*In his room high atop the Hotel Sidney Monad, Cyril
Zaki found his mind was aching. He had spent the last
five hours working on his explanation of why he needed
to spend five million dinars to refit the Top Dog's thruster
tubes on an uninsured planet. Zaki lit another melloroon
and tried to gather his thoughts. The repair platform
in orbit around Sidney had been surprisingly well
equipped and efficient for an obscure pleasure planet.
The maintenance foreman, who seemed accustomed to
spaceships in difficulty, was quite familiar with the Dog's
thrusters and confidently told Cyril that he could be on
his way after five days on Sidney. A gaily decorated
shuttle called the Wazoo took Zaki and his crew quickly
and smoothly down to the capital of Sidney: Ripov. But
instead of being reassured by the technical expertise
around him, Cyril was perturbed by the gaudy
advertisements that covered the walls of the shuttle's
lounge:*

> *Up close? Hotel Monad makes you relax!
> Who needs dinars? Gamble at Gotchas!
> Cancel your policy—zoroos live free!*

*The hotel room was relaxing after the whirling lights,
perfumed air, and shimmering hydroxyl fountains on*

the strips of Ripov. Evening light filtered through colored screens, a complicated sound-massage system swirled around the bed, and the videotape library had a fine collection of Cyprian's classics.

Cyril's work was interrupted near dusk by a tapping at the door.

"Cyril, are you in there?" asked Lisa Dalby.

Cyril strode across the room and opened the door. "Of course I am," he answered, a bit testily.

"Why don't you come out for a look at Sidney?"

"I'm trying to write a number two memo. I can't go out on the strips of Ripov and haul ashes with you and the crew."

"Cyril," Lisa countered, "You're just using that memo as an excuse. You're really up close because Sidney's uninsured."

Cyril inhaled on his melloroon, gagged slightly, and spoke again.

"Lisa, I'm a man of action, not words. Dictating this report is killing me; I've just got to finish it."

"Well, you'll never do it by smoking so many melloroons. Cyril, believe me, this planet's fantastic! It's got things you wouldn't believe! The city's full of gambling casinos where people actually risk losing dinars, there's pill palaces, 3-D hologram houses, saloons, dancing parlors, museums, ancient beautiful buildings called kurshes, and great consumption I don't know what else—and outside the city there's all sorts of parks with strange plants, weird animals, colored lakes—it's amazing! And there's no concrete anywhere."

Cyril's mind turned this over and he muttered, "Yeah, sure, but what about the zoroos and the other uninsured freaks? And what about the parts of the planet without premiums?"

"Well, frankly, Cyril, no areas seem to be insured, but nobody seems to mind! You've just got to see it! Julie and Bernie and I are going out for another look. As long as we stick together, there's no big risk. We'd like you to come along, Cyril; will you?"

"Well," Zaki replied, as he walked over to the ancient bed that dominated his room. It was a hydroxyl couch, and as he lowered himself toward it, warm and soothing ripples nudged his zor. Cyril looked at Dalby, at his feet, and at the mellow paneling on the walls of his room. *Lisa is right,* he thought, *I do need a break. And what can go wrong if we stick together?* "All right, Lisa," he said, "I'll do it."

He decided to follow his quick decision with rapid action.

"We might as well go right now, no sense wasting time. A little light amusement, then I can get this memo taped."

"Great," said Lisa. "Bernie and Julie are in the foyer; we can pick them up on the way."

Cyril Zaki's huge hands trembled slightly with exhilaration as the four explorers walked down the broad strips of Ripov. The contrasts were stunning: majestic ancient buildings, museums and storehouses, mingled with the slick shine of the amusement palaces. Zaki couldn't quite believe that people were standing in long lines waiting to crowd into the casinos to risk losing money. *What happens to these people when disaster strikes?* he wondered. *If this were evening back on Pilar, things would be a lot safer.* In fact, he reflected, not much would be going on aside from some careful computations of the safety factors. The streets would be quiet, calm, and deserted. Zaki noticed crowds of zoroos on the Ripov strips, all dressed in particolored sashes, apparently unworried about their lack of insurance. One of them was playing a tamboor and singing softly. Cyril wondered how Boyer was taking to the environment.

"Say," he asked, "Have you seen much of Boyer?"

Zots gave a small chuckle and answered, "No, but I'm sure he's enjoying himself." He pointed with a pudgy finger to the various zoroos and back to his ear. "Poor risks, eh?"

Lisa Dalby broke in, "Do we want culture and museums, or drinking and dancing?"

Zaki choked on his melloroon. "But drinking's illegal."

"Not on Sidney," replied Lisa.

Zaki hadn't been dancing in a long time, and the idea appeals to his natural sense of grace. And, well, he had drunk a little at the university, so he knew it wouldn't kill him. Besides, it was legal here. "Good risk, let's go drinking and dancing," he exclaimed with the same assurance he felt in knowing that he was the captain.

The rest of the group seemed pleased to follow this suggestion. Nearby beckoned a small, noisy, but pleasant-looking establishment, the Ever-Growing Balloon, and soon the four merrymakers found themselves in a corner booth overlooking the dancing area.

Cyril lit a melloroon with one hand, and with the other he hailed a barmaid. "Bring us four whatneys," he intoned as if he made a similar order every day.

Julie Card leaned toward Cyril and asked, "Do you see that small blond woman talking to the bartender and nodding in our direction. I'm sure she was in the hotel lobby earlier this evening. Does she look familiar to you?"

"No," replied Cyril, "but I . . ."

His answer was interrupted by the arrival of the drinks, for which Cyril let Zots pay. Lisa Dalby hoisted her beaker.

"Atoms to you," she toasted.

"Atoms to atoms," sounded around the table.

After the third round of whatneys, Cyril was feeling confident enough to ask, "Lisa, would you like to dunce, I mean dance?"

"Sure—what about you, Bernie and Julie?"

"No," Zots replied, "I'm feeling kind of tired. I think I'll just sit here a while and turn in to the music."

In the dance area, couples writhed gracefully in heavily cushioned sway-holes to the heavy beat. Zaki prided himself on his coordination, but he had a difficult time following the music, and the flashing lights made his mind reel. His thoughts seemed hazier than usual as he hummed along:

"The head bone connects to the leg bone, the leg bone connects to the back bone. . . ."

"Say, Lisa," Cyril shouted over the music, *"Let's go back to the booth, I'm feeling insecure."*

"Sure. I'm a bit wiped over myself."

When they returned to the corner, Cyril and Lisa saw that Julie and Bernie both had their heads on the table, apparently sleeping. Zaki's mind was working overtime to fight off his dizziness as he struggled to light a melloroon. He managed to put two and two together, came up with three, which was close enough, and said, *"I think we've been drugged, we're going to be uninsu . . ."*

Blackness engulfed him, and he slowly slid to the floor.

All matter is made of tiny bits and pieces called elementary particles. These elementary particles not only form all matter, but they are also responsible for all the forces between various kinds of matter in the universe. Even the sun's energy is a result of interactions among elementary particles. It is at the size level of elementary particles that Einstein's convertibility between mass and energy becomes a reality.

Elementary particles are much smaller than atoms, and hence they are invisible to the naked eye. Although they are extremely small and elusive, these particles affect all our lives; to ignore them would place us in the same position as people in the Middle Ages who ignored disease germs because they could not see them.

Because elementary particles are extremely small, physicists must use complicated and expensive equipment to study them. Giant particle accelerators that probe the world of elementary particles by smashing the particles into each other at large speeds have been built. Rather than use this expensive equipment, we shall employ our imagination in the search for elementary particles.

Consider a woman who looks at her hand with the ability to pretend that she is looking into a supermicroscope. With no magnification, she would simply see the usual features of her palm: her life and heart lines and so on. If she were to increase the magnification of her microscope 1,000 times, she would see the individual cells [they are about one-thousandth (10^{-3}) of a centimeter in diameter] that make up her skin. A further increase of 1,000 times in the magnifying power of her microscope would reveal the molecules out of which her skin cells are made [one-millionth (10^{-6}) of a centimeter in diameter]. By increasing the magnification 100 times more, she would be able to see the atoms [one-hundred-millionth (10^{-8}) of a centimeter in diameter] that form the molecules. For the woman to see what lies inside an atom, she would have to increase the power of her microscope

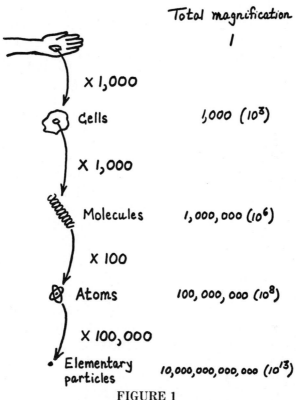

Total magnification

1

× 1,000

Cells   1,000 (10^3)

× 1,000

Molecules   1,000,000 (10^6)

× 100

Atoms   100,000,000 (10^8)

× 100,000

Elementary particles   10,000,000,000,000 (10^{13})

FIGURE 1
A comparison of the sizes of cells, molecules, atoms, and elementary particles.

an additional 100,000 (10^5) times. At this point the total magnifying power of her microscope would be 10 trillion (10^{13}) times. The world of elementary particles begins here, with the constituents of an atom.

Each *atom* contains three kinds of elementary particles: protons, neutrons, and electrons. Although the world contains fantastic numbers of atoms, their basic constituents are identical. All the protons in the universe are copies of one another; each neutron is like every other neutron, and every electron is equal to any other electron. The number of protons, neutrons, and electrons in a particular atom characterizes that kind of atom. Ninety-two different kinds of atoms occur naturally. Table 1 (page 38) lists some of the more common kinds of atoms, such as hydrogen, helium, carbon, and oxygen.

All atoms resemble the simplest atom: hydrogen. We may imagine a hydrogen atom as a tiny solar system formed by *one* electron orbiting around a center, or nucleus, of *one* proton (Fig. 2a)*; this is analogous to the earth orbiting around the sun (Fig. 2b).

The proton that forms the center of a hydrogen atom has a mass a little greater than one trillion-trillionth of a gram (a proton's mass is actually 1.6724×10^{-24} gm), and a proton has an approximate radius of one ten-trillionth (10^{-13}) of a centimeter. Protons are so small that it would take 10 trillion of them to span the width of a fingernail. Still smaller, an *electron* has a mass 1,836 times less than a proton's mass (the mass of an electron is 0.91×10^{-27} gm). For most purposes, an electron's radius can be considered much smaller than a proton's radius. In a hydrogen atom, the electron's orbit around the proton has a radius of five-billionths (5×10^{-9}) of a centimeter (Fig. 2a). We can construct a scale model of a hydrogen atom from familiar objects to illustrate the atom's structure: a grapefruit for the proton and an overweight ladybug for the electron. To maintain the proper scale, the ladybug must orbit the grapefruit at a distance of 5 miles (Fig. 3). This scale model reminds us that a hydrogen atom is mostly empty space. The same is true for all atoms and consequently for familiar forms of matter. Hence automobiles, smog, people, and peanut butter consist mostly of empty space.

We began a description of protons and electrons by giving their masses and approximate sizes. To understand atoms we must also deal with the electrical properties of protons and electrons. Although a proton and an electron have very different masses, they have electric

---

*More complicated atoms have more electrons in orbit around the nucleus.

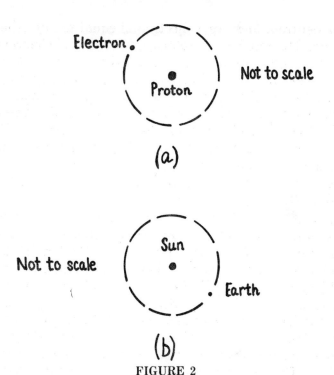

FIGURE 2

(a) A model for a hydrogen atom, showing the electron in orbit
around the proton. A hydrogen atom is held together by the
electromagnetic attraction between the proton and the electron.
This drawing is about 10^8 times larger than a hydrogen atom.
(b) A model for the earth's orbit around the sun in our solar system.
The earth and the sun attract each other through gravitational
forces. This drawing is about 10^{13} times smaller
than the solar system.

charges that are *equal in magnitude but opposite in sign.*
A proton has one unit of positive electric charge ($+e$);
an electron has one unit of negative electric charge ($-e$).
Ordinary electric current is usually just the slow movement of many individual electrons through a conducting
material. Quite generally, electric current is the flow of
electrically charged particles. All elementary particles
have electric charges that are a whole number times the
fundamental unit of electric charge, which we write as e.
Particles can have an electric charge of $+2e$ or $-7e$ or 0

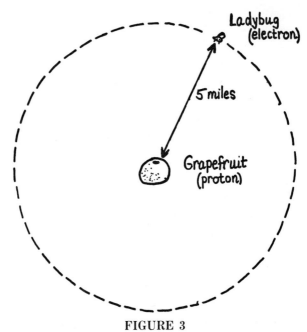

FIGURE 3
A scale model for a hydrogen atom can be made with
a grapefruit to represent the proton and a ladybug
to represent the electron. The ladybug must then
orbit the grapefruit at a distance of 5 miles.

but never electric charges like $+2.3e$ or $-1.5e$. Since al
matter is composed of elementary particles, it can have
only a whole number of units of electric charge.

One important feature of electrically charged particle:
is that they interact with each other through electro
magnetic forces. Electromagnetic forces *repel* any tw
particles with the *same sign* of electric charge away from
each other, and two particles with electric charges c
opposite sign are *attracted* to each other by electromag
netic forces (Fig. 4). Hydrogen atoms are held togethe
by the electromagnetic attraction between the positivel
charged proton and the negatively charged electron. Th

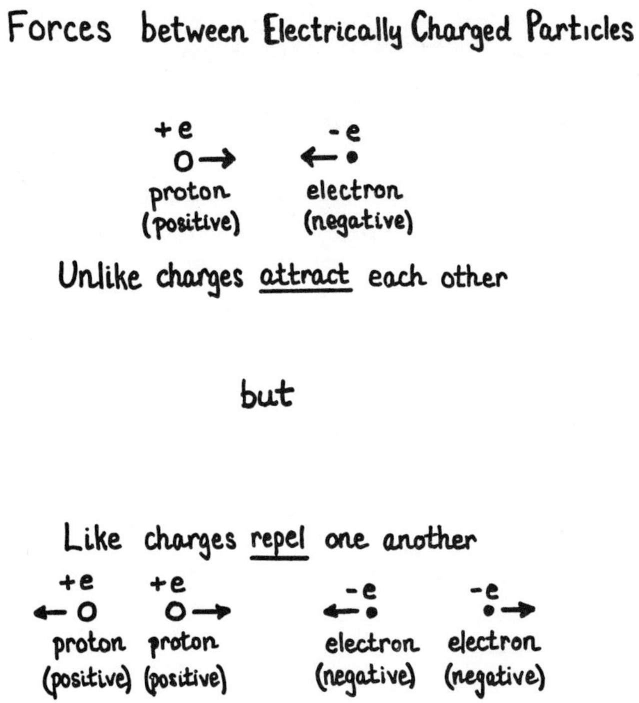

Unlike charges <u>attract</u> each other

but

Like charges <u>repel</u> one another

FIGURE 4

Electromagnetic forces between electrically charged particles
always *attract* particles with the opposite sign of electric charge
and *repel* particles with the same sign of electric charge.

total electric charge of a hydrogen atom, like that of all
atoms, is zero; that is, the proton's electric charge $+e$
plus the electron's electric charge $-e$ is zero.

Before we can describe other atoms we must introduce
another sort of elementary particle: the *neutron*. Each
neutron has a radius about the same as a proton's and a
mass that is only 0.2 percent larger than the mass of a
proton (the mass of a neutron is 1.6747×10^{-24} gm). However, a neutron is electrically neutral and has *no* electric
charge, hence the name neutron. Particles with no electric charge do not feel electromagnetic forces because
electromagnetic forces act only between particles with
some electric charge.

Protons, neutrons, and electrons are not the only types of elementary particles, but they are the three most common ones in the world, and indeed, by themselves these three kinds of particles appear to form most of the matter in the universe.

Now that we have introduced the neutron we can describe atoms other than hydrogen. Recall that a hydrogen atom, the simplest of atoms, consists of one electron (electric charge $-e$) in orbit around one proton (electric charge $+e$) and that the atom's total electric charge is zero. The next simplest atom is the helium atom. A helium atom's nucleus contains two protons (electric charge of two protons equals $+2e$) and two neutrons (electric charge of two neutrons equals zero) orbited by two electrons (electric charge of two electrons equals $-2e$). A helium atom's total electric charge is also zero. A helium atom, like a hydrogen atom, is held together by the electromagnetic attraction between the positively charged protons and the negatively charged electrons. We can construct a scale model for a helium atom by using four grapefruit for the nucleus and two ladybugs for the electrons: ordinary grapefruit represent protons and two pink grapefruit represent neutrons. To preserve the proper scale, the two ladybugs must orbit the four grapefruit at a distance of $2\frac{1}{2}$ miles (Fig. 5). The helium atom's nucleus would consist of the four grapefruit (two protons and two neutrons) pressed close to one another. This model illustrates again that helium atoms, like hydrogen atoms, are mainly empty space because they have just a tiny nucleus orbited at a large distance by two electrons.

All atoms are basically similar in structure to hydrogen and helium atoms. They all have a tiny nucleus composed of protons and neutrons, around which the electrons orbit at relatively enormous distances, and an atom always has a number of electrons equal to its number of protons, so the total electric charge of an atom is always zero. An atom that has gained or lost one or more elec-

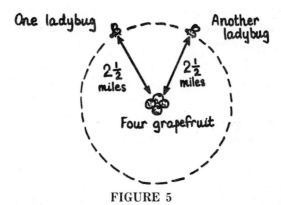

FIGURE 5

We can make a scale model for a helium atom by using
two ordinary grapefruit to represent the two protons,
two pink grapefruit to represent the two neutrons, and
two ladybugs to represent the two electrons. Then the ladybugs
must orbit the four grapefruit at a distance of $2\frac{1}{2}$ miles.

trons is called an *ion* and has a net electric charge. For
example, a sodium atom (Na) that loses one electron
becomes a sodium ion (Na⁺) with an electric charge of
+e. A chlorine atom (Cl) that gains one electron becomes
a chlorine ion (Cl⁻) with an electric charge of −e.

The different kinds of atoms are conveniently described
by two numerical quantities. The first quantity is called
the *atomic number*, which is the number of protons in
the atom's nucleus. The second quantity is the *atomic
weight*, which is the total number of protons *plus* neutrons in the atom's nucleus. Atoms with the same number
of protons but different numbers of neutrons in their
nucleus are called *isotopes* of one another. Isotopes have
the same atomic number but a different atomic weight.
For example, an ordinary helium atom has two protons
and two neutrons in its nucleus, so its atomic number is
2 and its atomic weight is 4. A rarer isotope of helium

has two protons but only one neutron in its nucleus, so it still has an atomic number of 2, but its atomic weight is 3 (see Table 1).

Most of the universe seems to consist of hydrogen (about 90 percent) and helium (about 9 percent). The next most abundant kinds of atoms are carbon, nitrogen, oxygen, and neon.

TABLE 1

Different kinds of atoms

Name	Atomic number (number of protons)	Number of neutrons	Atomic weight*
Hydrogen	1	0	1
Deuterium (heavy hydrogen)	1	1	2
Helium 3	2	1	3
Helium 4	2	2	4
Lithium 7	3	4	7
Carbon 12	6	6	12
Carbon 13	6	7	13
Carbon 14	6	8	14
Nitrogen 14	7	7	14
Oxygen 16	8	8	16
Fluorine 19	9	10	19
Neon 20	10	10	20
Sodium 23	11	12	23
Magnesium 24	12	12	24
Aluminum 27	13	14	27
Silicon 28	14	14	28
Chlorine 35	17	18	35
Iron 56	26	30	56
Arsenic 75	33	42	75
Silver 107	47	60	107
Tin 120	50	70	120
Platinum 195	78	117	195
Gold 197	79	118	197
Mercury 202	80	122	202
Lead 207	82	125	207
Uranium 235	92	143	235
Uranium 238	92	146	238

*The atomic weight equals the number of protons plus the number of neutrons.

The chemical properties of an atom describe how the **39**
atom can combine with other atoms and how large aggregates of atoms and molecules behave. For example, two hydrogen atoms and one oxygen atom can combine chemically to form a molecule of water. An atom's chemical properties are related to its number of electrons. The ability of atoms to bind to one another arises from the electromagnetic interaction of one atom's electrically charged constituents with another atom's electrically charged constituents. These electromagnetic interactions involve primarily the outer electrons of the various atoms, and they do not directly involve the tiny nuclei at the atoms' centers.

The number of electrons in an atom can in turn be related to its atomic number. We can understand this when we recall that an atom's number of electrons (each with electric charge $-e$) must equal the number of protons (each with electric charge $+e$) in the atom's nucleus for the total electric charge of the atom to be zero. Atoms with the same atomic number but different atomic weights still have the same chemical properties because they have the same number of protons and the same number of electrons. Thus atomic isotopes have the same chemical properties and differ from one another only in their nuclear structure.

The properties of the *nucleus* of an atom, as distinct from the chemical properties of the entire atom, can be described roughly by the atomic weight, which is the total number of neutrons plus protons in the atomic nucleus. An example of the difference of nuclear properties is uranium (refer to Table 1). Uranium atoms, which all have an atomic number of 92, have two isotopes that occur naturally on earth. The more common isotope, uranium 238 (U^{238}), has 146 neutrons and 92 protons in each atomic nucleus, so its atomic weight is 238. This uranium isotope is stable and does not tend to undergo nuclear disintegration (*nuclear fission*). The other uranium isotope, uranium 235 (U^{235}), has atoms with 143 neutrons and 92 protons in each nucleus, so the atomic weight of these atoms is 235. This uranium isotope does tend to disintegrate naturally by nuclear fission.

Atomic isotopes whose nuclei tend to disintegrate are called *radioactive*. Such radioactive isotopes provide a useful research tool in many fields. We can describe one example. Carbon has a radioactive isotope, carbon 14 (C^{14}), with an atomic weight of 14. Any molecule that contains carbon atoms can have carbon 14 atoms substituted for some of its ordinary carbon 12 (C^{12}) atoms. After this substitution, the molecule is said to be "doped." The substitution can occur because any two atomic isotopes have identical chemical properties even though they have different nuclear properties. Because of its radioactivity, the carbon 14 isotope enables us to trace the subsequent activity of a particular doped molecule or group of molecules. By using this tracing technique we have been able to unravel and understand many of the details of photosynthesis in plants and respiration in animals. These processes are fundamental for the continuance of life on earth.

From a human point of view, the most interesting kind of atom may well be carbon. A single carbon atom can combine chemically with as many as four other atoms. This is in contrast with other common kinds of atoms, such as nitrogen atoms, which can usually link up with three other atoms: oxygen, which can combine with two other atoms; hydrogen, which can combine with one other atom; and helium and neon atoms, which refuse to combine with any other atom at all.[1] The carbon atoms' readiness to combine with four other atoms, plus the fact that the connections can be made in four different directions, makes the linking of huge molecular chains possible. In particular, carbon is a key binding constituent of the giant molecules that carry genetic information in all living cells. These molecules are called DNA

---

[1]Because of their inability to combine with other atoms, helium and neon are called *inert gases*.

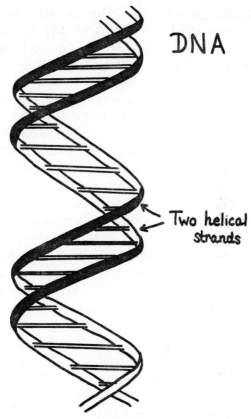

FIGURE 6

A schematic representation of a molecule of DNA
(deoxyribonucleic acid). Each DNA molecule has two long
helical strands that are linked together by short molecules.
Most of the thousands of individual atoms in a DNA molecule
are carbon, hydrogen, oxygen, or nitrogen.

(deoxyribonucleic acid) and are large, two-stranded, helical molecules composed of millions of atoms (Fig. 6). Carbon, hydrogen, oxygen, and nitrogen atoms are all present in these molecules. DNA molecules must be large and complex because the genetic information they transmit is complex. It seems likely that intelligent life on earth owes its emergence to the ever-increasing complexity of longer and longer molecules. Human bodies

42   consist mostly of water (80 percent) and large, carbon-based, "organic" molecules (20 percent). No one knows whether life systems based on the chemistry of carbon atoms are the only possible living organisms or whether the surfaces of planets like our earth are the only possible sites for the formation of such molecules. We may speculate, for instance, that certain clouds of interstellar gas with about the same temperature as the earth's surface might eventually form molecules complex enough to evolve an intelligence of their own. Also, silicon atoms (atomic weight 28) can also combine with four other atoms, but their bonds with the other atoms are weaker than those that carbon atoms can form, and silicon atoms are rarer (on a universal scale) than carbon atoms. It is possible, though, to imagine wholly different life forms based on the chemistry of silicon rather than carbon atoms.

Calculations of the exact way in which electrons move within an atom bring us into the realm of *quantum mechanics*. In general, quantum-mechanical calculations are needed to make an accurate description of systems of small (elementary) particles that are separated by small distances. By "small" we mean distances less than or equal to the radius of an atom, or about one hundred-millionth (10^{-8}) of a centimeter. At these small distances the results of quantum-mechanical calculations show that particles' behavior often differs from what we might have concluded from our everyday experiences with large objects like footballs, pizzas, and automobiles. An important example of this difference is the hydrogen atom.

We described a model for a hydrogen atom in which an electron orbits around a proton in a circle. Our daily experience tells us that the electron's orbit could have any size for its radius, ranging from zero to infinity. This seems natural because in an analogous situation, that of an artificial satellite orbiting around the earth, the satellite's orbit can have any size for its radius, such as 4,200 miles, 4,203.6 miles, 4,444 miles, and so forth.

In contrast, quantum-mechanical calculations show that in a hydrogen atom only certain, definite values are possible for the average distance of the electron from the proton.[1] In terms of our model of circular electron orbits, the only allowed values for the radius r of the orbit are those given by

$$r = N^2 \times (0.53 \times 10^{-8} \text{ cm})$$

where N is one of the integers, so $N = 1, 2, 3$, and so on. *No* other orbits are possible under the rules of quantum mechanics. The smallest orbit for the electron in a hydrogen atom is that for which $N = 1$ and $r = 0.53 \times 10^{-8}$ cm; the next smallest orbit has $N = 2$ and $r = 2.12 \times 10^{-8}$ cm, and so forth. In hydrogen atoms, and in other atoms as well, the electrons tend to seek the *smallest* possible orbit. That is, the electrons try to get as close to the nucleus as they can under the rules of quantum mechanics.

When we consider atoms with several electrons, we find that quantum-mechanical calculations also show that there is a *maximum number of electrons which can fit into a given orbit.* This rule is also in contrast with our experience with satellites because we can put as many satellites as we choose into any orbit. For a hydrogen atom we can characterize each allowed electron orbit by the quantum number N in the formula

$$r = N^2 \times (0.53 \times 10^{-8} \text{ cm})$$

For helium atoms, this quantum-mechanical formula is

$$r = N^2 \times (0.265 \times 10^{-8} \text{ cm})$$

---

[1] Quantum-mechanical calculations also show that we can not locate the electron with complete accuracy within an atom, so we must talk about the *average* distance of the electron from the proton throughout the orbit rather than an exact distance of the electron from the proton at every instant.

and in general, for atoms with atomic number equal to Z, r follows the formula

$$r = N^2 \times \frac{(0.53 \times 10^{-8} \text{ cm})}{Z}$$

Quantum-mechanical calculations show that for any atom the maximum number of electrons that can fit into an orbit characterized by a given N is $2 \times N^2$. Therefore,

A Model for a Neon Atom

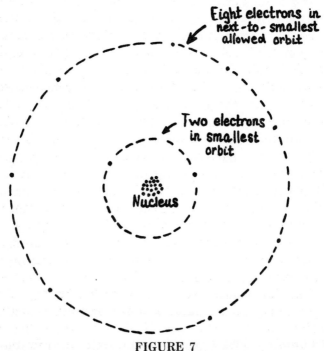

FIGURE 7

A model for a neon atom, showing the 10 electrons in orbit around the nucleus. There are 10 protons and 10 neutrons in the nucleus of a neon atom. The two electrons in the smallest orbit are the maximum allowed for that orbit and make the orbit a closed shell. Likewise, the eight electrons in the next smallest orbit are the maximum allowed for *that* orbit and make it a closed shell too.

two electrons can fit into the smallest orbit ($N = 1$), eight electrons can fit into the next smallest orbit ($N = 2$), and so on. Orbits with the maximum number of allowed electrons in them are called *closed shells*. For example, the 2 electrons in a helium atom make the smallest orbit a closed shell, and the 10 electrons in a neon atom (Fig. 7) make the two smallest orbits closed shells. Atoms like neon and helium, whose electrons are all orbiting in closed shells, almost never combine with other atoms; they are chemically *inert*.

We can understand an atom's chemical properties in terms of its quantum-mechanical orbits or shells; we shall consider a few examples. Lithium has an atomic number of 3. The first two electrons in a lithium atom close the $N = 1$ orbit or shell. The third electron occupies the $N = 2$ orbit, which can hold a maximum of eight electrons (Fig. 8). In a chemical reaction an atom tries to reach a situation with other atoms where only closed shells exist. Thus to form a molecule, lithium atoms try to donate their outer atom in the $N = 2$ shell to another atom so that only the closed $N = 1$ shell is left. On the other hand, the lithium atom could accept seven more electrons and thus close its $N = 2$ shell completely. It is usually easier for a lithium atom to give away *one* electron than to accept *seven* new electrons, so a lithium atom combines with other atoms by giving away its one outer electron. A lithium atom that donates one electron has a chemical *valence*, or worth, of $+1$. If lithium tended to accept seven electrons, it would have a valence of -7. Note that donation of electrons gets a plus sign and acceptance of electrons gets a minus sign in the valence.

Consider another example: fluorine. Fluorine has an atomic number of 9, so each fluorine atom has nine electrons. The first two electrons in a fluorine atom close the $N = 1$ shell, and the next seven electrons occupy the $N = 2$ orbit (Fig. 9). To form a chemical compound, a fluorine atom accepts one electron from another atom to close the $N = 2$ shell, and in this situation fluorine has a valence of -1. (It is highly unlikely that fluorine would

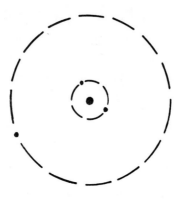

Lithium (Li)

One electron in N=2 shell

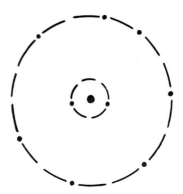

Fluorine (F)

Seven electrons in N=2 shell

FIGURE 8

A lithium (Li) atom ($Z = 3$) tends to donate its one electron in the $N = 2$ orbit to another atom during the formation of a chemical (ionic) bond. In doing this it is left with a closed $N = 1$ shell or orbit. A fluorine (F) atom ($Z = 9$) tends to accept one electron to fill its $N = 2$ orbit during the formation of a chemical (ionic) bond.

lose all seven of its outer electrons, thus acquiring valence of +7, in order to be left with just the $N =$ closed shell.)

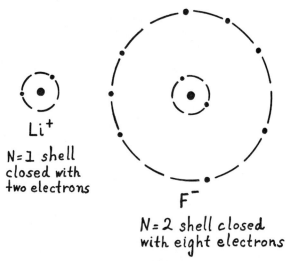

Li⁺

N=1 shell
closed with
two electrons

F⁻

N=2 shell closed
with eight electrons

FIGURE 9
Li-F is held together by the net positive charge
of the Li⁺ (ion) and the net negative charge of the F⁻ (ion)
attracting one another through electromagnetic forces.

In a chemical reaction where lithium (Li) and fluorine (F) combine to form poisonous lithium fluoride (LiF), the lithium donates the one electron in its $N=2$ shell to the fluorine atom, which uses the electron to fill its $N=2$ shell. The closure of both the $N=1$ shell in the lithium atom and the $N=2$ shell in the fluorine atom makes lithium fluoride a very stable molecule. Molecules in which the positive valence of one constituent atom just matches the negative valence of the other are almost always extremely stable.

Still another example is carbon, which has an atomic number of 6. In carbon atoms the two inner electrons close the $N=1$ orbit, and the four outer electrons occupy the $N=2$ orbit, which has room for eight electrons in all. In a chemical reaction that involves carbon atoms, several possibilities can occur. The carbon atom can lose its four outer electrons and thus be left with just a closed $N=1$ shell. In such a situation, the carbon atom would have a valence of +4. The carbon atom can also accept four electrons to fill its $N=2$ shell; the atom would have a valence of −4. Both possibilities exist because they involve the same change in the number of outer electrons. However, the usual way that a carbon atom links with other atoms is by sharing some or all of its four outer electrons with other atoms in what is called a *covalent bond*. The great flexibility in the type of bonds that carbon atoms can form with other atoms makes carbon a fundamental element in the building of complex organic molecules; thus carbon atoms are found at the heart of all organic processes.

The strange behavior of elementary particles also appears when we measure the *masses* of atomic nuclei. We find that elementary particles (protons and neutrons) bound into an atomic nucleus do *not* have the same mass they would if they were separated by distances much larger than the size of a nucleus. For example, a nucleus of hydrogen 2 (H^2), which is called a *deuteron*, consists of one proton and one neutron. Taken separately, a proton has a mass of 1.6724×10^{-24} gm and a neutron has a mass of 1.6747×10^{-24} gm. However, the mass of a *deuteron* is a bit less than the sum of these two masses; instead of having a mass of $1.6724 \times 10^{-24} + 1.6747 \times 10^{-24} = 3.3471 \times 10^{-24}$ gm, the deuteron nucleus has a mass of 3.3432×10^{-24} gm. This sort of decrease in mass is a common characteristic of protons and neutrons bound together into atomic nuclei. We may think of the decrease as resulting from the energy needed to hold the nucleus together. The need to provide this binding energy saps the particles' essence a tiny bit and causes the total mass of a nucleus to be a little less than the sum of the masses of the protons and neutrons that form the nucleus.

SUMMARY

Atoms are made of elementary particles called protons, neutrons, and electrons. All atoms have electrons orbiting around nuclei made of much heavier (more massive) protons and neutrons. The tiny nuclei fill only a tiny fraction of the total volume of the atom. Within an atom the positively charged protons exert an attractive electromagnetic force on the negatively charged electrons. Neutrons have no electric charge and experience no electromagnetic forces. The attractive electromagnetic force holds the atoms together. The number of electrons in an atom always equals the number of protons in the atom's nucleus. Different *kinds* of atoms are distinguished by the number of protons and electrons they have, into different *elements* (hydrogen, helium, carbon, and so forth). If two atoms have the same number of protons and electrons but different numbers of neutrons in their nucleus, they have the same chemical properties and are said to be isotopes of each other.

Quantum-mechanical calculations show that for the electrons in orbit around an atom's nucleus, only certain, definite orbits are allowed, each characterized by a certain integer N. The maximum number of electrons that can fit into a given allowed orbit is $2 \times N^2$. In any atom, the electrons tend to seek the orbit with the smallest possible value of N.

QUESTIONS

1 What three kinds of elementary particles make up all atoms?
2 Which has more mass, a proton or an electron? By how much?
3 Do two protons repel or attract each other by electromagnetic forces?
4 Does the electromagnetic force between two electrons repel or attract them?
5 A carbon atom has six protons and six neutrons in its nucleus. How many electrons does it have in orbit around the nucleus?

6 Suppose that one of the electrons was ripped off a helium atom. What would be the total electric charge of the remainder?

7 Atoms with one or more of their electrons gone (ripped off) are called ions. Would it be fair to say that the proton is a hydrogen ion?

8 Are the chemical properties of the isotopes carbon 12 and carbon 13 different? (Table 1 on page 38 lists some elements and their isotopes.)

9 What is the diameter of a hydrogen atom? About how many hydrogen atoms would it take to cover a line 1 cm long? (*Ans.*: 9.43×10^7 atoms)

10 Given that only $2 \times N^2$ electrons can fit into the atomic orbit characterized by N ($N = 1, 2, 3$, and so on) and that electrons tend to seek the orbits with the smallest possible N, how many electrons should be in the $N = 3$ orbit in an atom of magnesium 24? (See Table 1.) (*Ans.*: 2)

11 Consider a model for atoms using grapefruit and ladybugs, such as described in Figs. 3 and 5. The smallest orbit ($N = 1$) for the electron in a hydrogen orbit then has a radius of 5 miles. Using the same scale model, how large would the radius be of the outermost electron in the magnesium atom described in Prob. 10? (*Ans.*: $3\frac{3}{4}$ miles)

12 What is the valence of an oxygen atom? (*Ans.*: -2) In the model described in Prob. 11, how many miles would the electrons in an oxygen atom in the $N = 1$ orbit be from the nucleus? How far for the $N = 2$ orbit? (*Ans.*: $\frac{5}{8}$ mile; $2\frac{1}{4}$ miles)

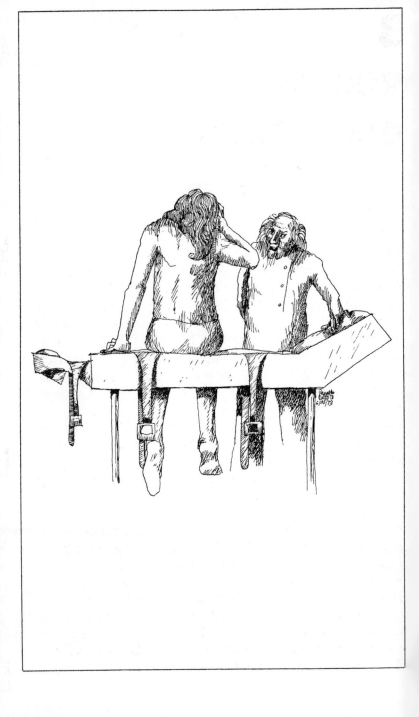

3
energy
and matter

Deep in the heart of a gigantic gray building, a
motionless Cyril Zaki lay strapped to a table. As he
climbed from the mists of unconsciousness, a ballet of
pain and nausea greeted him. The stray ends of his huge
body seemed to have grown new nerves since he was last
awake. Each of them determined to dominate the others
by producing a tingle of pain in his neck, a long rolling
spasm in his solar plexus, or a surge of dizziness. His
head felt like an overripe pumpkin. In an effort to orient
himself, Cyril tried to examine his surroundings, but
the fire in his brain kept his eyes from focusing. His left
hand made a feeble gesture to light a nonexistent
melloroon in a vain attempt to comfort himself.

   A soft voice from behind his head asked, "Feeling
insecure, Captain?"

   Zaki, startled, strained against the straps, only to
have tiny stars, giant squid, and mashed potatoes
explode in his head. Defeated for the moment, he relaxed
into the foamy covering of the table, letting the last
needle of pain disengage itself from his cortex. Vague
flashes of the evening with Zots, Card, and Dalby
appeared and disappeared. Black holes ringed by long
worms twisted and burned in his brain. Was there a
connection between the black holes and the sway-holes?
Zaki remembered dancing to the throbbing beat in the

54 *sway-holes, and the drinking. This drinking is a bum voyage, Cyril thought, no wonder it's illegal. I feel totally tubed out. He felt a pinprick in his right bicep and realized that his bonds had been loosened.*

"Health and security," said the same soft voice behind him.

Cyril sat up a little too quickly and felt new streams of nausea wash over him. Sharp pains peppered his abdomen with right hooks, left jabs, and a few low blows. When his eyes worked once more he saw a small, grey-haired man with no eyebrows holding a sort of pliers and wearing a look of immense benevolence.

"I'm Dr. Zed," the gnomish man said, "Bokomoru P. Zed. Please don't feel worried, insecure as you would call it."

Zaki felt worried. He felt totally insecure.

"What's the come-off here?" he asked breathily.

"You were feeling bad, Captain," said Dr. Zed. You needed help and I gave it to you. For instance, that injection you just had. In a few minutes, it should remove the headache and nausea you're presently experiencing."

Zed waddled over to a stainless-steel bin that protruded from the wall and dropped his plierlike tool into it. Zaki, who was already feeling physically improved, followed Zed successfully with his eyes, and as he did so he realized that he lay inside an immense room with no windows, one steel door, and a high domed ceiling. Along the walls were units of technical equipment not found on any of the space vessels Cyril knew, and he knew them all. There was an aura of the past about some of the glittering tools of steel and plastic that occupied the room, and this puzzled Zaki because everything showed signs of recent use. None of the equipment bore the red and gold linked arms of the Guaranteed Trust insurance seal. One entire wall was given to videotape recesses, but the neatly lettered titles were strange, even to Zaki: Jules Caesar. Candy. Brave New World. Eightfold Way. Einstein. S-matrix Theory.

in bold red capitals: PERMIDS. Zaki mused over the
titles; some of them joggled his memory. Aha, that's it,
he thought, and the mantra of the zoroos surfaced in
his brain:

> *Weinstein/rama/Einstein/karma*
> *Matter/energy/ever warmer*
> *Swinging/ringing/don't do thinging*
> *Matter/energy/keep on singing*
> *Weinstein/dharma/Einstein/karma*
> *Energy makes matter warmer*

But weren't Einstein and Weinstein just the heroes of
space fables? Something that Zaki couldn't fathom was
terribly wrong, and he felt totally lost, uninsured. But
this was wrong; he was insured. His premiums were
paid in full. Zaki thought of his past, of the slogans that
had guided him through puberty:

> *An insured man is a proud man.*
> *The only trip a zoroo can make is the big trip.*
> *Nothing's too small to save.*
> *Girls want the man with the big premium.*

Thoughts of his big premium gave Cyril some self-
confidence, and in a tone of authority modeled on a
melloroon commercial he asked defiantly:
"Where am I?"
"All in good time, my boy," half answered Dr. Zed.
"I'll be right with you."
Zed was sorting through a plastic hamper in search
of something, so Cyril took advantage of the moment to
twist around and examine the room behind him. More
familiar objects appeared, consoles and computers, with
their flashing lights and rows of computo-touch buttons.
Seated at one of the consoles, manipulating the manifold
dials, was a vaguely familiar blond woman dressed in a
purple lab coat. Zaki continued to feel stronger physically

but mentally more bewildered. Where was he? What was he doing here? Where were his friends, chubby Bernie Zots, slim Lisa Dalby, competent Julie Card?

"Here, my boy, I found these for you," said the slight Dr. Zed, as he offered Zaki a crushed package of melloroons.

When Cyril reached for them, he noticed for the first time that he was sleeveless. In fact, he was pantless too, since his magnificent body was now housed in a drab single-piece pullover.

"Bad risk," he exclaimed, "Where are my clothes?"

"I had my assistant remove them," Zed chuckled. "I thought you'd be more comfortable, and a little hesitant about leaving us too soon. You know, before we had a chat."

Zaki's first feeling of humiliation at the hands of the blond assistant quickly gave way to the more solid taste of fear. He lit a melloroon and inhaled deeply.

"Cough, choke, gasp," Cyril rasped, "These are stale!"

"Well, we can't have everything," beamed Dr. Zed. "But you are the Captain."

Cyril managed to smile back at his host, but he wondered if B. P. Zed and his equipment were insured at all. He'd never dealt with a man who had no eyebrows and no premiums.

The squat Dr. Zed seemed to read Zaki's mind. He asked:

"Tell me, Captain Zaki, have you ever wondered why everything must be insured?"

"How did you know my name?" asked Cyril as he took a cautious hit off his melloroon.

"I know quite a bit about you, Captain. For example, are you not Zenith Borg's top investigator? But that's neither here nor there. Come now, haven't you wondered why we need all that insurance?"

"Of course I have," Cyril replied. What a dum-dum, he thought. "No insurance, no security. No security, no society. No society, no health. Do you think we all ought to become a pack of zoroos? And where would they be if

we didn't have insurance to bail them out of the messes **57**
they blunder into?"

"Very pretty, Captain. Quite the phrasemaker, you are.
I suppose sailing through space brings out the poet in
you. But let me ask you this: Why do you also insure all
those crazy items you don't know the first thing about?
Whose health do they protect?"

Bokomoru's point rubbed Zaki the wrong way,
especially as he had no good answer with which to put
down the garrulous doctor.

"It seems safest to insure whatever we can," he said.
"Of course, there are difficulties with assessment, and
with replacement if something gets destroyed. But that's
far better than not having any insurance at all, isn't it?"

Dr. Zed pressed on unheeding. "Now look, Cyril—may
I call you Cyril?" Zaki gave a sort of belch of approval.
"Do you know that there was a time when men sailed
into space without any insurance at all?"

Zaki had heard vague stories along these lines from
the crazies who blasted their minds at macrocosmic
dope stores. And there were the bedtime spacetales and
legends that implied the same thing, but Cyril knew
they were only old myths. Why, some of them even
insisted that human beings had all started on the
same planet.

"I don't believe it," he said. "And if it were true, what
difference would that make? They say everyone used to
die before they were even a hundred years old—so what?"

"But, Cyril," Zed murmured in a conspiratorial tone,
"Suppose that you were free to venture into space fearless
and uninsured. What if you knew the true significance
of that black sphere on your ship's bridge? Suppose you
could unlock the mystery of the Permids*?"*

Zed peered at Zaki as if expecting a definite response
to his last words, but Cyril maintained a muscular pout.
That word again, he thought, and he sent nerve impulses
coursing through the adyts of his brain, trying to
remember where he had heard it before. Unsuccessful,
he started talking.

58   "But I am free," Cyril said. "I can travel through space anywhere, provided I take out the proper premiums. After all, that's the advantage of insurance; we spread the dangers so everyone shares the risks."

"Spare me the rest of the Guaranteed Trust creed, if you please," said Zed. "I'm perfectly familiar with it. 'Insure while you have insurability'; 'Keep the wolf of disability from the lamb of society'—what crap! Believe me, Cyril, what you're doing is destroying the noblest impulses human beings have. You're part of the reason that people are forgetting things faster than they learn them."

"Who says they are?" Cyril demanded truculently, but even as he spoke he realized that Zed was right about people getting dumber.

"Don't waste my time with these false gestures of scorn, Captain. You know in your heart that you're twice as incompetent as your grandfather was. Insurance as a way of life has simply proved to be a failure. So long as enough people keep coming up with new ideas and ways to meet new problems, it makes perfectly good sense to have the risks spread around. But once everyone believes that society rests on having everyone protected, once even the best minds spend their time guarding and insuring their own zors, there's nowhere to go but downhill. Your constipated, souvenir-ridden society saves everything it can't reproduce, without knowing why. Can't you see that we have beautiful green parks here on Sidney, while all the insured planets are covered with crowding, polluting useless, junk?"

Zaki recoiled at the obscenity "junk," but he was relieved to learn that he was still on Sidney. Immediately thoughts of escape began to bubble inside him. Zed rambled on:

"You see, Cyril, the old legends were correct. Men did venture into space without insurance. Because in the pre-preem days men and women had the ability to build and replace equipment; they didn't need insurance to *cover their fears. They were brighter and tougher than the monkeys we've become."*

"Monkeys my zor," Cyril rebutted. He felt that if he could keep Zed talking, he could work out some escape plan. "How do you know that's true?"

Zed seemed to enjoy the sound of his voice; with its soft s's and r's, he sounded like a sports announcer. In a tone of well-oiled satisfaction he rolled on: "I haven't always been on Sidney, my friend. I spent many years on Pilar, working for the Trust. Does that surprise you, my boy?"

"Well, er, ah, yes," mumbled Cyril Zaki. Given the right circumstances, even his premium could surprise him.

"Oh, why ask? — even your premium could hold untold mystery for you. Yes, at gargantuan Trust headquarters, I worked in the canceled policies office, and had to use many ancient languages. You don't seem to know what it means, but I had to sift right down to the Permids. Day after day, as I buried my nose in the past . . . ."

"Heh heh, giggle," snorted Zaki, as he tried to imagine Zed's nose in the past.

"What's that, my boy?"

"Oh, nothing, I was just thinking."

"What a refreshing change — may you have beginner's luck. But as I was saying, I probed the past, rooting through the canceled policies, and I kept meeting more and more references to what we carelessly call myths, like the Eightfold Way and the S matrix. Gradually, watching old videotapes and ancient books, I found . . ."

"Books?" Zaki exhaled a cloud of grey melloroon smoke.

"Yes, that's it, books. Before videotapes, people wrote thousands of words on paper to store information; primitive but workable. Now where was I? Ah, yes, with each bit of new information I began to realize that those ancient myths and spacetales had a lot of the truth in them. All those years in the labyrinthine basements of

the Trust will pay off for me yet. Even then I knew that the man who unraveled those legends would have ultimate power. He could build and unbuild things. Transmute them through space and time. Energy to matter, matter to energy — orgone energy!"

Zed paused briefly, panting like a zoroo. "You see, Cyril, the age of knowledge really did happen."

A cold sweat broke out behind Zaki's dimpled knees. He lit another melloroon. "Agh, cough, gasp, houma," he said. "How come I never heard about this at the Trust?"

"Well, my boy, when I told my superior about these investigations, she laughed at me. Told me, B. P. Zed, that my tubes were out of line."

"I can't understand it," intoned Cyril gravely, producing his most sympathetic expression. Got to humor this heretic, he thought, my kitzle depends on it. Zed's mobile face took on the glow of a father telling his son about dinars, and he told Zaki more and more about his life at the Trust.

"Finally my friends and colleagues came to ridicule me," he said. "They issued a 16-B memo to keep me from circulating "heresy," and they removed me from the canceled policies office!"

"A 16-B memo," mouthed Cyril reverently, sticking a thumb in his ear.

"But the Trust will regret it!" Zed turned with outstretched hands encompassing the huge room as he spoke. "Here I am rediscovering the power of old. A few more pieces, and when I have them it will be the junk-pile for those stopped-up snivelers. We'll show the whole G.T. crowd a thing or two, you can bet your boogaloo on that, my overmuscled friend . . . ."

Zed's sibilant solo voice boomed out at Zaki, seeming to fill the giant room, leaving no place for Cyril's escape. I've got to get back to Pilar and stop this madman, he thought. Trust only know what he can do. After all, he captured me, C. V. Zaki. But how to divert his attention and get him to leave me alone? Cyril spoke up.

"Who is this 'we' you're talking about?"

"Aha, my boy, don't think I didn't recognize you as a spy as soon as you landed. Just how dumb do you think I am? Let me tell you a bit more about what I've accomplished already. Why, on Werner it was I who found the long-buried tapes of the great Wheeling Stones . . ."

Zed's narrative was interrupted by his blond assistant, who rose from her computer console and crossed the room to him. The rhythmic motions of her perfect zor set Zaki's premium to rising, temporarily blotting out his thoughts of escape. The assistant placed a golden arm around the doctor's plump shoulders and whispered into his shell-like ear. Zed swallowed noisily and spoke with a lopsided smile.

"Excuse us, Captain, but my assistant has reminded me of some crucial business that demands critical attention. We'll return shortly; meanwhile just try to relax, if you know what that means."

Kiss my relax, Cyril thought to himself, but spoke no reply.

Once we have penetrated the atom and seen that it is formed of constituents much smaller than the atom itself, namely protons, neutrons, and electrons, it is natural to ask: "How can we penetrate the world of elementary particles like the proton and discover its secrets?" Physicists have found that the best way is to shake the particles very hard and see what happens. If we were tiny enough, and a tough enough detective like Sam Spade in "The Maltese Falcon," we could simply grab a proton by its lapels and shake it until it revealed the information we wanted. But because we are not this small we must use more subtle techniques to shake protons.

This shaking process can be accomplished on a microscopic level by arranging the collision of one elementary particle with another elementary particle. After the collision, we can study the resulting elementary particle debris. This method of obtaining information is like smashing two cars together and then examining the bits and pieces that are spewn out in an attempt to find out what cars are made of (Fig. 1).

To study particle collisions and what is produced (shaken out) in them, physicists have used a few billion tax dollars to build huge elementary-particle accelerators, such as the Stanford Linear Accelerator. These

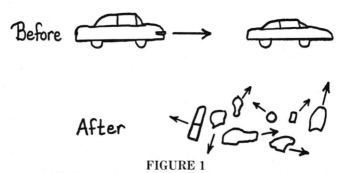

FIGURE 1
We can smash cars into one another and then examine the bits and pieces in an attempt to find out what cars are made of.

machines use electromagnetic forces to accelerate charged elementary particles to high velocities and then smash them into one another.

Particle accelerators are not the only places where collisions between elementary particles can occur. Inside stars like our sun, elementary particles (mainly protons) undergo many collisions with one another. The most important result of these collisions for us is the conversion of part of the protons' mass into energy that eventually reaches us on earth in the form of light and heat. Even inside our own bodies, collisions of elementary particles can occur with good, bad, or zero effects. "Cosmic rays" made mainly of fast-moving protons continually arrive on earth from deep space and eventually pass through our bodies without doing any damage. In cancer therapy, elementary particles are sometimes made to pass through cancerous tissue, where they collide with the elementary particles in the diseased matter and ultimately destroy it. On the other hand, nuclear explosions or nuclear power-plant accidents can produce such a huge outflow of elementary particles that their collision with the particles in a person's body would be fatal.

Let us ignore, for the moment, the huge man-made accelerating machines and the insides of stars to concentrate on the particles themselves. It is in the realm of elementary particles that we find Einstein's principle of the convertibility of matter and energy to be of great importance.

When two rapidly moving elementary particles (protons, for example) collide, *new* matter (particles) is sometimes created out of their energy of motion. Alternatively, some of the mass of the colliding particles can sometimes be transformed into additional energy of motion.

Consider one proton (the target) at rest and another proton approaching it (Fig. 2). Sometimes new particles, in addition to the original two protons, will appear after the collision of two such protons. To illustrate the creation and convertibility of matter and to introduce three

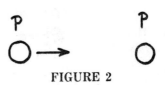

FIGURE 2

more elementary particles, we shall describe two typical proton-proton collision reactions.

In the first example, two protons collide, and afterward, in addition to the original two protons, two new particles called neutral pi mesons (π^0) are created (Fig. 3). This is an example of Einstein's principle of the creation of mass out of the energy of motion. What happens is that some of the energy of motion of the colliding particles is converted into the energy of mass of the additional particles that are produced. (At the end of this chapter, we discuss in detail the convertibility between mass and energy and show how to use Einstein's famous equation $E = M \times c^2$.)

Not only can new matter be created, but matter can also be *transformed* into other forms. We can illustrate this by describing another possible outcome of a collision between two protons. Sometimes after two protons collide, they are transformed into two neutrons, and in addition two positively charged pi mesons (π^+) are created (Fig. 4).

Because of the statistical (or quantum-mechanical) nature of elementary particles it is impossible to predict

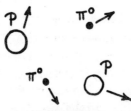

FIGURE 3

After the collision of two protons, we sometimes find not only two protons but also two neutral pi mesons (symbol π^0).

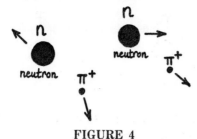

FIGURE 4
After the collision of two protons, we sometimes find
two neutrons and two positive pi mesons (symbol π^+).

just what will happen each time two protons collide. The
same initial situation can have quite different results. If
we study 100 proton-proton collisions, we can only say,
for example, that 10 percent of the time two neutral pi
mesons will be created or 20 percent of the time two positive pi mesons will be created; in the other 70 percent of
the cases, other reactions will occur. The important point
is that *all* the different kinds of elementary particles can
be created from the appropriate collisions of one kind of
particle with another.

Pi mesons are, on the average, 6.8 times *less* massive
than protons, and they come in three varieties, with
positive electric charge $(+e)$, negative electric charge
$(-e)$, and no electric charge. The positive pi mesons and
the negative pi mesons (π^-) have equal mass $(0.2488$
$\times 10^{-24}$ gm), but the neutral pi mesons have slightly less
mass $(0.2405 \times 10^{-24}$ gm) than the charged pi mesons.
All three kinds of pi mesons have approximately the
same radius as a proton (about 10^{-13} cm).

Protons, neutrons, pi mesons, and electrons are not the
only elementary particles that exist in our universe and
are relevant to our environment. We have used them as
convenient examples of the creation and convertibility
of the energy of motion and the energy of mass and as
illustrations of how new elementary particles can appear.

TABLE 1
Some kinds of elementary particles

Particle	Symbol	Mass, gm	Electric charge
Proton	p	1.6724×10^{-24}	$+e$
Neutron	n	1.6747×10^{-24}	0
Electron	e^-	0.00091×10^{-24}	$-e$
Positive pi meson (pi-plus meson)	π^+	0.2488×10^{-24}	$+e$
Negative pi-meson (pi-minus meson)	π^-	0.2488×10^{-24}	$-e$
Neutral pi meson (pi-zero meson)	π^0	0.2405×10^{-24}	0

Table 1 lists the particles that we have mentioned so far, together with their masses and electric charges.

All the different kinds of elementary particles can be created from collisions among various other elementary particles. The convertibility of motion into particles with mass, and of one kind of particle into another kind, lies at the root of the basic processes in the universe.

We can now study the conversion of the energy of motion into mass, and, alternatively, the conversion of mass into the energy of motion. We shall use two concepts: (1) Einstein's mass-energy relationship and (2) the principle of *conservation of energy*.

Every particle has an energy associated with its mass, called its *energy of mass*. Einstein's mass-energy relationship states that the energy of mass is proportional to the mass of the particle times a constant number. In Einstein's famous equation, the energy of mass is

Energy of mass = (mass of the particle) × (speed of light)²

The constant of proportionality between the particle's mass and its energy of mass is the speed of light squared. This famous relationship applies to every particle that we know. The more massive a particle is, the greater its energy of mass. Einstein's relationship may be written in a more compact form as

$$E = M(\text{particle}) \times c^2$$

where E is the energy of mass, $M(\text{particle})$ is the particle's mass, and c is the speed of light, a fundamental constant of nature (c is almost exactly 300,000 kilometers/sec, or 3×10^{10} cm/sec, which equals 186,000 miles/sec).

Consider a single proton. The proton's energy of mass is

$$\text{Energy of mass of proton} = M(\text{proton}) \times c^2$$

For a *system* of particles, the energy of mass is the sum of the individual particles' energies of mass. For example, consider the system that consists of a proton and a neutral pi meson (Fig. 5); the energy of mass of the system is

$$\text{Energy of mass} = M(\text{proton}) \times c^2 + M(\text{pi meson}) \times c^2$$

Every known particle or system of particles has an energy of mass that is given by the sum of each particle's energy of mass. In addition, if a particle is moving, the particle has an energy of motion, called its kinetic energy (KE). If there are no forces acting on the particle, then the particle's *total energy* is the sum of its energy of mass and its kinetic energy. A single proton has a total energy of

$$\text{Total energy (proton)} = M(\text{proton}) \times c^2 + (\text{KE})_{\text{proton}}$$

FIGURE 5

The total energy of a system of particles, if no forces are acting on it, is the sum of each particle's energy of mass and kinetic energy;[1] that is, the system's total energy is the sum of each particle's total energy. For example, consider the system of one proton and one neutral pi meson drawn in Fig. 5; here the total energy is

$$\text{Total energy} = \overbrace{M(\text{proton}) \times c^2 + M(\text{pi meson}) \times c^2}^{\text{system's energy of mass}}$$
$$+ \underbrace{(KE)_{\text{proton}} + (KE)_{\text{pi meson}}}_{\text{system's kinetic energy}}$$

The key to understanding what happens when elementary particles collide is Einstein's mass-energy relationship combined with the principle of *conservation of energy*. The principle of conservation of energy states that in any process the *total energy remains the same*. This is true even if the form of the energy changes. At the beginning of any process there is a certain amount of total energy, and at the end there is an identical amount of total energy. This principle may be summarized as

Total energy after = total energy before

To understand and to feel the power of Einstein's mass-energy relationship and the conservation-of-energy principle, we shall look at two examples. First, we consider a collision of two protons that produces two neutral pi mesons in addition to the original two protons. Some of the protons' energy of motion before the collision is

---

[1] If M is the proton's mass and v its velocity, the proton's kinetic energy is

$$(KE)_{\text{proton}} = \frac{M \times c^2}{\sqrt{1 - (v^2/c^2)}} - (M \times c^2)$$

If v^2 is much less than c^2, this formula has the approximation

$$(KE)_{\text{proton}} \simeq \tfrac{1}{2}M \times v^2 \qquad \text{if } v^2 \ll c^2$$

transformed into the neutral pi mesons' energy of mass **69** (Figs. 2 and 3).

Consider the two protons before they collide. One proton is moving rapidly toward the other, stationary proton (Fig. 2). Each proton has an energy of mass given by Einstein's relationship as $M(\text{proton}) \times c^2$. Since there are two protons, their total energy of mass is $2 \times M(\text{proton}) \times c^2$. In addition, because one of the protons is moving there is an energy of motion or kinetic energy associated with it, which we shall call $(\text{KE})_{\text{before}}$. The total energy of the system before the collision is therefore given by the sum of $2 \times M(\text{proton}) \times c^2$ plus $(\text{KE})_{\text{before}}$, or

$$\text{Total energy before} = 2 \times M(\text{proton}) \times c^2 + (\text{KE})_{\text{before}}$$

After the collision, besides the two protons there are two new particles: the neutral pi mesons. Some of the initial energy of motion or kinetic energy before has been converted into the pi mesons' energy of mass (Fig. 3). Since the two protons and the two pi mesons may be in motion after the collision, they can have kinetic energy, which we shall call $(\text{KE})_{\text{after}}$. This is the sum of the kinetic energies of each of the four final particles. The total energy after the collision is the sum of each particle's energy of mass *plus* its kinetic energy, so

$$\text{Total energy after} = \overbrace{\begin{array}{l} M(\text{proton}) \times c^2 + M(\text{proton}) \times c^2 \\ + M(\text{pi meson}) \times c^2 + M(\text{pi meson}) \times c^2 \end{array}}^{2 \times M(\text{proton}) \times c^2 + 2 \times M(\text{pi meson}) \times c^2}$$
$$\left.\begin{array}{l} + (\text{KE})_{\text{proton 1}} + (\text{KE})_{\text{proton 2}} \\ + (\text{KE})_{\text{pi meson 1}} + (\text{KE})_{\text{pi meson 2}} \end{array}\right\} (\text{KE})_{\text{after}}$$

The last four terms together make up the kinetic energy after $[(\text{KE})_{\text{after}}]$, so we have

$$\text{Total energy after} = 2 \times M(\text{proton}) \times c^2$$
$$+ 2 \times M(\text{pi meson}) \times c^2 + (\text{KE})_{\text{after}}$$

The *total energy* is the same before and after the collision process, so the kinetic energy after must be less than the kinetic energy before to compensate for the increase in the energy of mass after the collision. We can neatly summarize all these words by using our energy-conservation equation:

$$\text{Total energy after} = \text{total energy before}$$

After we substitute the expression for the total energies before and after, we find that the conservation-of-energy equation tells us that

$$2 \times M(\text{proton}) \times c^2 + (KE)_{\text{before}} = 2 \times M(\text{proton}) \times c^2$$
$$+ 2 \times M(\text{pi meson}) \times c^2$$
$$+ (KE)_{\text{after}}$$

This equation can be solved to give the result

$$(KE)_{\text{before}} - (KE)_{\text{after}} = 2 \times M(\text{pi meson}) \times c^2$$

which tells us exactly how much energy of motion was transformed into the energy of rest mass because of the collision process. The additional energy of rest mass is that of the two neutral pi mesons which were created.

As a second example of Einstein's mass-energy relationship and the conservation-of-energy principle, we consider how the sun generates its energy. The sun converts some of its energy of mass into energy of motion. This energy of motion eventually reaches us in the form of light and heat. This energy-conversion process occurs through a series of collisions among elementary particles. The series is called the *proton-proton cycle* because the first step of the cycle involves the collision of two protons. We shall consider the first step of the proton-proton cycle in detail to show how some of the protons' energy of mass can be transformed into energy of motion.

Consider two protons about to collide (Fig. 2). The total energy of mass of the system is $2 \times M(\text{proton}) \times c^2$. Also, because the protons are moving (for generality, we now assume that both protons are in motion before they collide), each proton has some kinetic energy. The protons must have a speed sufficient to overcome their mutual electromagnetic repulsion before they can collide. We can add the two protons' individual energies of motion and call the result $(\text{KE})_{\text{before}}$ for the system of two protons. The total energy of the system before the collision is therefore

Total energy before = energy of mass before

+ kinetic energy before

$= 2 \times M(\text{proton}) \times c^2 \times (\text{KE})_{\text{before}}$

Whatever happens during the collision, the total energy after must be the same number as the total energy before.

When the proton-proton cycle occurs, we find after the collision of two protons a deuteron d, a positron e^+, and a neutrino v_e; the result is shown in Fig. 6.

A deuteron has one unit of positive charge and a mass of 3.3432×10^{-24} gm; it is the nucleus of the atomic isotope hydrogen 2 that we described at the end of Chap. 2. A positron has the same mass as an electron, but it has one unit of *positive* electric charge. A neutrino has *no* mass and *no* electric charge, but it *can* have energy of motion.[1]

When we add the masses of the three particles produced by the collision between the two protons, we find that the total mass after the collision is *less* than the total mass before the collision (Table 2). This means that some energy of mass of the initial two protons has been converted into energy of motion as a result of the collision. We may consider this energy of motion as "liberated" (changed into energy of motion from energy of mass) during the collision process.

---

[1]We shall discuss positrons and neutrinos in more detail in Chap. 5.

After

FIGURE 6
The collision of two protons can produce a deuteron (d),
a positron (e^+), and a neutrino (ν_e). These final particles
have more kinetic energy than did the two initial protons.

TABLE 2
Conservation of energy in the proton-proton cycle (step 1).
Total energy before = total energy after.

Before

Mass of initial particles:

$$M(\text{proton}) = 1.6724 \times 10^{-24} \text{ gm}$$

$$M(\text{proton}) = 1.6724 \times 10^{-24} \text{ gm}$$

Mass before $= 3.3448 \times 10^{-24}$ gm

Energy of mass before $= (3.3448 \times 10^{-24} \text{ gm}) \times c^2$

Kinetic energy before $= (\text{KE})_{\text{before}}$

Total energy before $= (3.3448 \times 10^{-24} \text{ gm}) \times c^2 + (\text{KE})_{\text{before}}$

After

Mass of final particles:

$$M(\text{deuteron}) = 3.3432 \times 10^{-24} \text{ gm}$$
$$M(\text{positron}) = 0.0009 \times 10^{-24} \text{ gm}$$
$$M(\text{neutrino}) = 0 \text{ gm}$$

Mass after $= 3.3441 \times 10^{-24}$ gm

Energy of mass after $= (3.3441 \times 10^{-24} \text{ gm}) \times c^2$

Kinetic energy after $= (\text{KE})_{\text{after}}$

Total energy after $= (3.3441 \times 10^{-24} \text{ gm}) \times c^2 + (\text{KE})_{\text{after}}$

Since the total energy before = total energy after, we find **73**

$$(KE)_{after} - (KE)_{before} = (3.3448 \times 10^{-24} \text{ gm}) \times c^2$$
$$- (3.3441 \times 10^{-24} \text{ gm}) \times c^2$$
$$= (0.0007 \times 10^{-24} \text{ gm}) \times c^2$$
$$= (7 \times 10^{-28} \text{ gm}) \times (3 \times 10^{10} \text{ cm/sec})^2$$
$$= 6.3 \times 10^{-7} \text{ erg}$$

To calculate how much energy of motion is liberated in the first step of the proton-proton cycle, we write the total energy after the collision as

$$\text{Total energy after} = M(\text{deuteron}) \times c^2 + M(\text{positron}) \times c^2$$
$$+ M(\text{neutrino}) \times c^2$$
$$+ \underbrace{(KE)_{\text{deuteron}} + (KE)_{\text{positron}} + (KE)_{\text{neutrino}}}_{(KE)_{after}}$$

Next we use the conservation-of-energy principle to write

$$\text{Total energy before} = \text{total energy after}$$

We find that

$$2 \times M(\text{proton}) \times c^2 + (KE)_{before}$$
$$= M(\text{deuteron}) \times c^2 \times M(\text{positron}) \times c^2 + M(\text{neutrino}) \times c^2$$
$$+ (KE)_{after}$$

Rearranging terms, we find

$$(KE)_{after} - (KE)_{before} = 2 \times M(\text{proton}) \times c^2 - M(\text{deuteron}) \times c^2$$
$$- M(\text{positron}) \times c^2 - M(\text{neutrino}) \times c^2$$

When we substitute the numerical values for the masses (see Table 2), we find that the increase in kinetic energy is

$$(KE)_{after} - (KE)_{before} = (3.3448 \times 10^{-24} \text{ gm}) \times c^2$$
$$- (3.3441 \times 10^{-24} \text{ gm}) \times c^2$$

$$= (0.0007 \times 10^{-24} \text{ gm}) \times c^2$$
$$= (0.0007 \times 10^{-24} \text{ gm}) \times (3 \times 10^{10} \text{ cm/sec})^2$$
$$= (7 \times 10^{-28} \text{ gm}) \times (9 \times 10^{20} \text{ cm}^2/\text{sec}^2)$$
$$= 6.3 \times 10^{-7} \text{ gm} \times \text{cm}^2/\text{sec}^2$$
$$= 6.3 \times 10^{-7} \text{ erg}$$

One erg, by definition, is an amount of energy equal to one gram of mass times (one centimeter per second of velocity) squared. This is about the amount of energy a housefly uses in taking off from a windowpane.

The amount of kinetic energy that is liberated when two protons collide and produce a deuteron, positron, and neutrino is less than one-millionth of an erg. Although this is a tiny amount of energy, the sun can produce great amounts of energy because huge numbers of these reactions occur every second. In each of these reactions, a small amount of energy of mass is converted into energy of motion. This energy of motion eventually reaches us in the form of light and heat, and thus the sun gives us its true warmth.

We have described the first step of the proton-proton cycle that liberates kinetic energy inside the sun. This step converts about 0.02 percent of the energy of mass of the two protons into energy of motion. Within the sun, there are about 2×10^{38} of these reactions each second.

The proton-proton cycle has two more steps in addition to the original collision of two protons. In the second step of the cycle, a proton and a deuteron collide, producing a nucleus of helium 3 (He^3), plus a photon (light ray). Figure 7 shows this collision process.

An He^3 nucleus consists of two protons and one neutron (see Table 1, Chap. 2). This collision process also liberates energy of motion from energy of mass; the additional energy of motion, 9×10^{-6} erg, is carried off mainly by the light ray.

The third step in the proton-proton cycle is the collision of two nuclei of He^3 to produce a nucleus of He^4 and two more protons (Fig. 8).

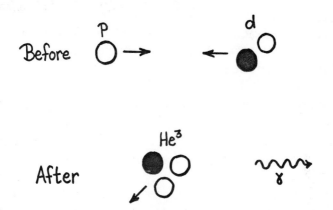

FIGURE 7
The collision of a proton (*p*) and a deuteron (*d*) can produce
a nucleus of helium 3 (He³), a light ray (γ), and some
additional energy of motion out of energy of mass.

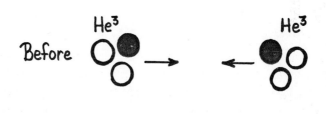

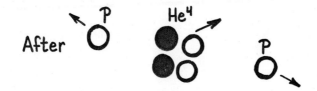

FIGURE 8
Two nuclei of helium 3 collide to produce one nucleus
of helium 4 (He⁴), two protons (*p*), and some additional
energy of motion out of energy of mass.

Like the first two steps of the proton-proton cycle, this collision process also liberates energy of motion from energy of mass: The three final particles have a combined energy of motion that is greater than the energy of motion of the two colliding He^3 nuclei by 2×10^{-5} erg.

Figure 9 summarizes the three steps of the proton-proton cycle. The net result of these three steps is to liberate an amount of kinetic energy equal to only 4

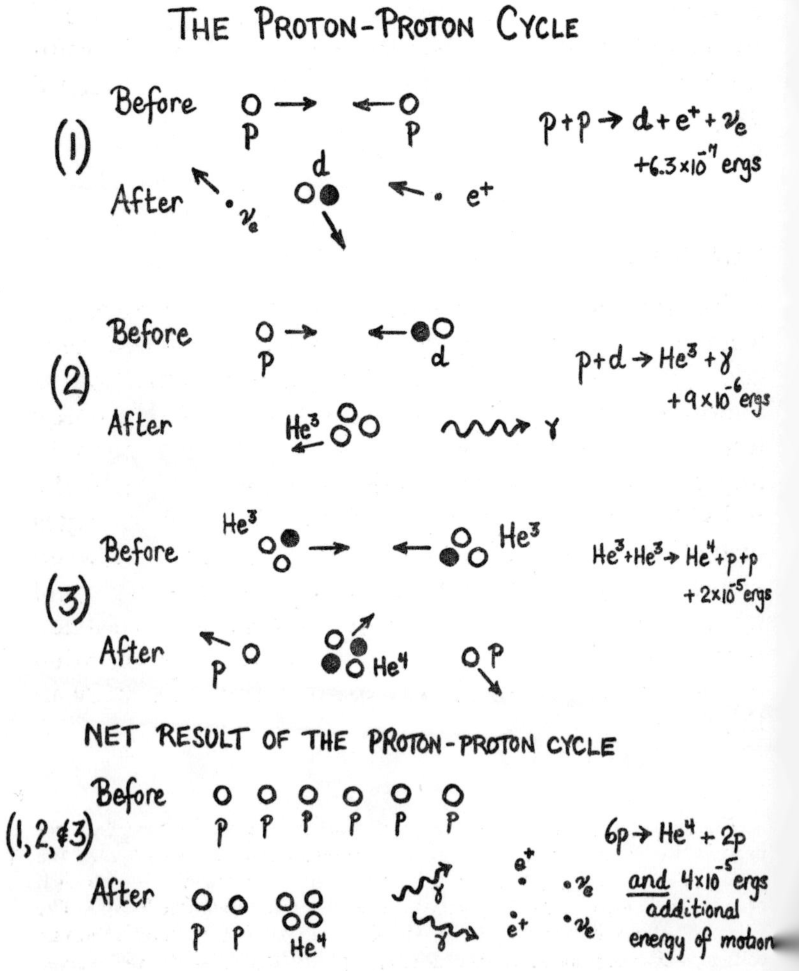
THE PROTON-PROTON CYCLE
(1)
Before
p
p
After
d
e+
$p+p \to d+e^+ + \nu_e$
$+6.3\times10^{-7}$ ergs
(2)
Before
p
d
After
He^3
γ
$p+d \to He^3+\gamma$
$+9\times10^{-6}$ ergs
(3)
Before
He^3
He^3
After
p
He^4
p
$He^3+He^3 \to He^4+p+p$
$+2\times10^{-5}$ ergs
NET RESULT OF THE PROTON-PROTON CYCLE
(1,2,&3)
Before
p p p p p p
After
p p
He^4
$6p \to He^4 + 2p$
and 4×10^{-5} ergs
additional
energy of motion

FIGURE 9

percent of a proton's energy of mass: 4×10^{-5} erg. Nonetheless, the proton-proton cycle produces enough energy of motion to keep the sun and most other stars shining for billions of years. The additional energy of motion produced in the proton-proton cycle appears as an agitation of the particles in the sun, that is, as an increase in the *temperature*[1] of the particles at the sun's center, where the proton-proton cycle is going on. By continual collisions among the sun's elementary particles, the energy of motion liberated in its interior passes to the particles at its surface, where the high temperature of the gas particles causes the energy of motion to be radiated into space at just the same rate as it is produced deep inside the sun.

SUMMARY

Collisions between elementary particles can change one kind of particle into another. In addition, such collisions can change energy of mass into energy of motion or energy of motion into energy of mass. The *total* energy (energy of mass plus energy of motion) stays the same in any process. To find the total energy, we must add the individual particles' energies of mass and their energies of motion. Inside stars like the sun, collisions between elementary particles constantly change some energy of mass into energy of motion through a series of collision reactions called the proton-proton cycle. This liberated energy of motion eventually escapes from the stars in the form of light and heat.

---

[1]The temperature of a group of particles measures the average energy per particle. Inside the sun, the temperature increases toward the center. At the center of the sun, the temperature is so high that protons collide with each other with enough energy to produce the cycle we have described above. The liberated energy of motion is shared with the other particles in the sun through collisions that are not so violent as to change the kind of particles involved in these collisions.

1   Is a proton bigger or smaller than an atom?
2   After a rapidly moving proton hits another proton, can other particles besides the original two protons appear?
3   Suppose that a man has twice as much mass as his wife. Who has more *energy* of mass? By how much?
4   How many times the energy of mass of an electron is the energy of mass of a proton? (*Ans.*: 1,836 times)
5   How much energy of mass does a woman have if her mass is 50 kg (50,000 gm)? If you express this answer in ergs, how does it compare with the energy used each year by people on earth, about 3×10^{27} ergs? (*Ans.*: 4.5×10^{25} ergs)
6   Consider a motionless neutron. Is it correct to say that *all* of this neutron's total energy is energy of mass?
7   Suppose that a rapidly moving proton collides with a stationary proton, and afterward there are two protons in motion. Has the total energy (energy of mass plus energy of motion) of the first proton remained constant in this collision process? Has the kinetic energy of proton one plus proton two remained the same during this process?
8   What are the three reactions that occur during the proton-proton cycle? At the beginning of each reaction, is there more mass than at the end of the reaction? What happens to this mass? Where is the most common place for these reactions to occur?
9   Consider the second step of the proton-proton cycle (see page 76). A proton and a deuteron collide to produce a nucleus of He^3. The masses of the deuteron and He^3 nucleus can be expressed as

$$M(\text{deuteron}) = 2 \times M(\text{proton}) - 1.82 \times M(\text{electron})$$
$$M(\text{He}^3) = 3 \times M(\text{proton}) - 10.5 \times M(\text{electron})$$

In terms of the proton and electron masses, how much total mass is present before the collision? How much total mass. is there after the collision? How much energy of motion is liberated? [*Ans.*: $3 \times M(\text{proton}) - 1.82 \times M(\text{electron})$; $3 \times M(\text{proton}) - 10.5 \times M(\text{electron})$; $8.68 \times M(\text{electron}) \times c^2 = 7.1 \times 10^{-6}$ erg]

4
forces

On Pilar the terminal — squat, dark, and formidable —
loomed behind the city like a mountain. Sentries moved
in a smooth shuffle around its exterior, silent persuaders
behind a ring of bold signs that warned Noninsurors to
keep out. Inside, a thousand corridors hummed with a
small army of Trust workers who dealt with the ongoing
affairs of the galactic arm. The Trust's employees
combined dedication with discipline, eager but fearful.
No detail remained beyond scrutiny, no stone unturned
in the quest for security. Far below ground level, deep
in the best-guarded bowels of the gigantic concrete
enclosure, the last functioning 9900 computer hummed
quietly. This mainstay of the insurance network digested
four trillion bits of information every minute, to spew
forth probabilities, inferences, risks, premiums,
potentialities. The computer was pampered by the Trust's
best cybernicians and insured for fifty billion dinars.

In the building's innermost sanctum on the O ring,
Zenith Borg sat behind a massive desk and ran her
small, supple hands over a teleconsole, giving the orders
that radiated to the far corners of the Trust's domain.
Myriads of star systems awaited her commands, although
it was highly irritating to Borg that her messages
traveled only at the speed of light and took years to reach
their destinations. Zenith idly wondered why the clods

82 *whom the Trust certified as scientists claimed there was no way to do better.*

Twenty years of directing the power of Guaranteed Trust had left grey threads in Zenith's long brown hair, but her face and eyes were unchanged from the day when she had proven to the Committee of Concomitance that state security demanded her elevation to the post of Insuresse Extraordinaire. And even now, so far as the computer could tell, there had been no mistake because the Trust's fortunes had been good. Nothing beyond the routinely unforeseen had ruffled the galactic ether, until the starships began to disappear out by Omicron Ceti.

Borg pressed several keys on the console, and the Cyril Zaki dossier flashed into view. Long experience let the I.E. pierce the mass of data quickly to reach the real nitty gritty. Zaki was rated AA: dedicated, trusting, rarely irascible, methodical, tough. Zenith wondered whether the top psyexperts would ever learn to include imagination in the list of qualities they assayed. Still, there was no doubt that Zaki had been the best man to send on his mission. But had he succeeded? Zenith Borg began to think about the human mass, to sift through her years of influencing seven hundred billion people. As a young woman she had felt some twinges of doubt over the manipulations she performed, but as she gained experience she realized that men functioned best under orders. Now, not even the trusted Zaki had been told why he was sent out as her personal probe of the cosmo. Borg knew that she must not risk her own person on missions like this, not because she had any fear, certainly not, but because she was too valuable to Guaranteed Trust. The I.E. was above petty fears and criticism. She must think for all womankind.

Zenith half closed her lustrous grey eyes and stroked the teleconsole. Power moved through her to the console and back again, real power, not that physical stuff Cyril was always bragging about. Borg wondered what it would be like to possess the ultimate power.

She had learned very early in life that there are only **83**
two important questions:

What do you want?

And what will you settle for?

*Ninety-nine point nine percent of humanity devoted
their energies to the first question, if they thought that
far. They daydreamed their way toward happiness, or
found ways to prove that life was mistreating them.
A lucky few, like Cyril Zaki, found themselves in tune
with their dreams. It might be one in a million who
managed to deal with the universe as it is, not as desires
project it, Zenith thought. Of these a few would settle for
nothing less than real power. Not too surprising we've
ended in control, Borg thought.*

*Her mind returned to a man who seemed to have
slipped out of these categories: Bokomoru Zed. What sort
of a bad risk was he? What if he didn't show his fine
hand when Zaki arrived? Why had Zed taken over
Sidney—for its memories?*

When Mick Jagger swaggers toward the microphone he exudes a sort of life force. The forces he exerts on his audience are a complicated set of psychological and physiological interactions that can not be simply understood. In the movie "Doctor Strangelove," General Jack D. Ripper tells his flight commander: "Mandrake, women sense my power." This sort of power is also difficult to understand. The forces and powers that determine the behavior of the world's politicians, military leaders, and normal people are extremely complex. In fact, these forces are so complicated that they are different for each person on earth.

In contrast, the physical forces that hold the universe together are quite simple to describe. There are only four forces: gravitational, electromagnetic, strong, and weak. Each kind of interaction or force plays a particular role in the balance of the universe, and each has its greatest effect only over certain distances.

For astronomical distances, gravitational forces dominate. As we shrink toward the world of the atom, electromagnetic forces become the most important. If we shrink to sizes smaller than atoms, down to the nucleus of an atom, we find that strong forces dominate the scene, and at such distances and even smaller ones weak forces will appear. We shall discuss, in turn, each kind of force and its key role in the preservation of the structure of the universe.

GRAVITATIONAL FORCES

Gravity is always an attractive force and has its most important effects between large bodies separated by large distances. For example, the motions of the stars in space are determined by gravitational forces. The gravitational force of attraction between the sun and the earth causes the earth to orbit around the sun (Fig. 1). The same sort of gravitational forces also causes the moon to orbit the earth. People themselves are held to the surface

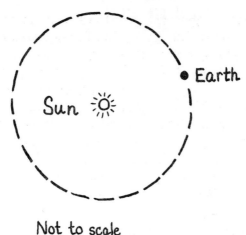

Not to scale

FIGURE 1

The earth orbits the sun under the influence of the sun's gravity.

of the earth by the force of gravity. And the motions of stars in space are determined by gravitational forces between them. We see, then, that gravity acts over large distances. The range of interaction of the gravitational forces we have just described is from 4,000 miles (the radius of the earth) to the gigantic distances between the stars.

All elementary particles are attracted to each other by the force of gravity. This gravitational force is very weak between any two elementary particles, so it takes huge aggregates of these particles, such as the earth, to create a noticeable gravitational force. Gravity becomes significant only at large distances because at short distances other forces are stronger and mask the effects of gravity. (By "large" distances we mean those larger than a few miles.) The fact that gravitational forces always *add* as we increase the number of particles is what allows gravitation to dominate situations with large distances and large numbers of particles.

Electromagnetic forces act only between electrically charged systems. They bind atoms into molecules and also hold groups of molecules together. Thus even huge groups of molecules such as Cyril Zaki, Lassie, and Mack trucks are held together by electromagnetic forces. Electromagnetic forces are attractive between particles with electric charges of opposite sign, and they are repulsive between particles with electric charges of the same sign. Particles with no electric charge experience no electromagnetic forces.

In atoms like the hydrogen atom drawn in Fig. 2, the electromagnetic attraction of the positively charged protons in the atom's nucleus for the distant, negatively charged electrons causes the electrons to orbit around the nucleus and thus holds the atoms together. This is analogous, but on a much smaller scale, to the sun's *gravitational* attraction for the earth, which causes the earth to orbit the sun. Electromagnetic forces are most

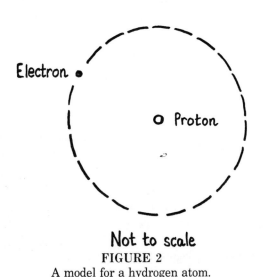

Not to scale
FIGURE 2
A model for a hydrogen atom.

effective — and dominate all other forces — at such distances as the size of an atom (that is, 10^{-8} cm or one hundred-millionth of a centimeter). If we make a model for a hydrogen atom in which a grapefruit represents the proton, then the electron could be a ladybug orbiting the grapefruit at a distance of 5 miles. Using the grapefruit model, we would say that electromagnetic forces have their greatest effect at distances of $\frac{1}{10}$ mile to 1 million miles; this corresponds to true distances of 10^{-10} to 10^{-3} cm.

STRONG FORCES

Strong forces hold together the nuclei of atoms. They are most effective at distances that are several times a proton's radius, 10^{-13} cm or one ten-trillionth of a centimeter. At these distances the strong forces completely overshadow all the other forces. For example, the nucleus of a helium 4 (He⁴) atom consists of two protons and two neutrons. These four particles are bound by strong forces (Fig. 3). Because both the two protons in the nucleus have positive electric charge, the electromagnetic forces between the protons tend to make them repel one another. Such electromagnetic forces would blow apart the helium nucleus (and indeed all the nuclei in the universe) were it not for the fact that at nuclear distances

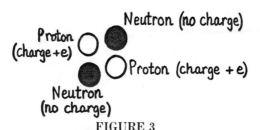

Helium Nucleus

FIGURE 3

The nucleus of a helium atom (He⁴) consists of two protons and two neutrons held together by strong forces. Strong forces hold together all atomic nuclei.

strong forces dominate the electromagnetic forces and hold the nuclei together. At distances several times greater than the radius of a proton, the strength of the strong forces decreases rapidly to zero. Strong forces are essentially zero for distances greater than a few times a proton's radius, about 10^{-13} cm. Within these distances, however, strong forces keep the protons and neutrons in atomic nuclei together. In terms of our grapefruit model, strong forces dominate all other forces at distances comparable to the diameter of a grapefruit. Thus the four grapefruit that represent the two protons and two neutrons in a helium nucleus are held together by strong forces.

WEAK FORCES

Weak forces are not true forces in the traditional sense; the term is used by physicists to describe a class of reactions that can occur between elementary particles. Weak reactions are a necessary part of the proton-proton cycle that liberates energy inside stars like our sun. We do not understand weak reactions very well, but we know that they occur only when particles are extremely close together (grapefruit distances or less); some evidence suggests that weak reactions occur at distances even less than the radius of a proton.

An ordinary neutron is one example of a weak reaction. If a neutron is isolated from an atomic nucleus and is far from any other elementary particles, it will decay after about 1,000 sec into a proton, an electron, and a massless, chargeless particle called an antineutrino (Fig. 4). This decay is produced by weak forces and is called a weak reaction. *Whenever antineutrinos or their close relatives, neutrinos, appear, we have a weak reaction.*

Let us summarize, in decreasing ranges of action, the four types of forces in the universe:

1 Gravitational forces act between all particles, but they are most important over astronomical distances. These forces determine the motions of celestial bodies.

Before   After

neutron ● n

p   ○   proton
e⁻   •   electron
ν̄ₑ   •   antineutrino

FIGURE 4

An isolated neutron decays through a weak reaction into a
proton, electron, and antineutrino after about 1,000 sec.

2   Electromagnetic forces dominate over distances of
atomic size and hold atoms together. They also bind
atoms into molecular chains, such as trees and people.

3   Strong forces hold atomic nuclei together and act
over distances about the size of a proton's radius (one
ten-trillionth of a centimeter).

4   Weak forces are a necessary part of the chain of
reactions that liberate energy in stars. Their range of
interaction is no greater than that of strong forces, and
may be less. What we call "nuclear reactions," such as
the steps of the proton-proton cycle discussed in Chap. 3,
are usually a combination of strong and weak reactions
between elementary particles.

Figure 5 summarizes the four types of forces and the
distances for which they are dominant.

Now that we have discussed the distances at which the
four types of forces act, we can compare directly the
strengths of two different kinds of force, namely, gravitational and electromagnetic. We shall find why the
electromagnetic forces dominate on the atomic scale of
distances (10^{-8} cm), and why gravitational forces dominate over large distances like the distance between the

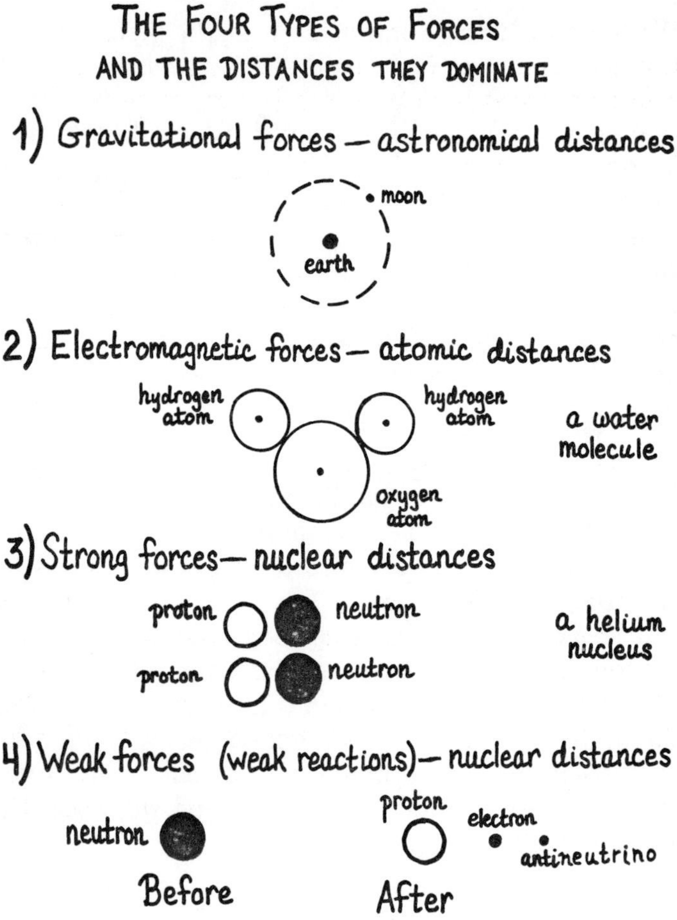

THE FOUR TYPES OF FORCES
AND THE DISTANCES THEY DOMINATE

1) Gravitational forces — astronomical distances

2) Electromagnetic forces — atomic distances

a water molecule

3) Strong forces — nuclear distances

a helium nucleus

4) Weak forces (weak reactions) — nuclear distances

FIGURE 5

earth and the moon (240,000 miles). We can directly compare gravitational and electromagnetic forces (we shall denote them here by $\mathscr{G}$ and $\mathscr{E}$, respectively) because they both have force laws. Recall that Newton's law of gravitation gives the force of gravity between any two bodies as

$$\mathscr{G} = -\frac{G \times (\text{mass of one body}) \times (\text{mass of other body})}{(\text{distance between centers})^2}$$

We have included a minus sign in the force to indicate
that the force is attractive. (By long-standing convention,
a force between two particles is attractive if it is negative and repulsive if it is positive.) Electromagnetic
forces obey a law similar to Newton's law of gravitation:
Coulomb's law. Coulomb's law of electromagnetism states
that the electromagnetic force between two electrically
charged particles acts along the line between their
centers, and the force is directly proportional to the
product of the particles' electric charges and is inversely
proportional to the *square of the distance between their
centers.* That is,

$$\mathcal{E} = \frac{K \times \text{(charge on one particle)} \times \text{(charge on other particle)}}{\text{(distance between centers)}^2}$$

K is a constant of proportionality, analogous to G in
Newton's law, and it is always the same number, everywhere. The units of the constant K are those that change
the $\text{(charge)}^2/\text{(distance)}^2$ on the right-hand side of the
equation into units of force. The basic unit of charge is
the *coulomb* (coul), named in honor of the man who discovered Coulomb's law. The electric charge of a proton
is $e = 1.6 \times 10^{-19}$ coul. If we measure charges in coulombs,
the constant K in Coulomb's law has the value

$$K = 9 \times 10^{18} \frac{\text{gm} \times \text{cm}^3}{\text{coul}^2 \times \text{sec}^2}$$

Electromagnetic forces attract two particles if they
have *opposite* signs of electric charge, and they repel
particles if they have the *same* sign of electric charge.
Coulomb's law for electromagnetic forces is analogous to
the law of gravitation, with the replacement of the constant G by K, the (mass of one body) by (charge of one
particle) and the (mass of other body) by (charge of other
particle). However, both kinds of forces *decrease as the
square of the distance* between the particles' centers.
We can now show why electromagnetic forces completely dominate gravitational forces inside an atom by

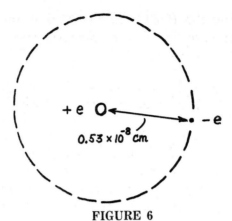

FIGURE 6
In a hydrogen atom, the distance between the proton's center
and the electron's center is 0.53×10^{-8} cm.

using the model for a hydrogen atom, where the electron
orbits the proton in a circle with a radius of 0.53×10^{-8} cm
(Fig. 6). First, we write the *electromagnetic* force of attraction of the proton for the electron, and then we write the
gravitational force of attraction of the proton for the
electron, so that we can compare the two forces. The
electromagnetic force between proton and electron is

$$\mathscr{E} = \frac{K \times (\text{charge of proton}) \times (\text{charge of electron})}{(0.53 \times 10^{-8} \text{ cm})^2}$$

and the gravitational force between proton and electron is

$$\mathscr{G} = -\frac{G \times (\text{mass of proton}) \times (\text{mass of electron})}{(0.53 \times 10^{-8} \text{ cm})^2}$$

To see how many times larger the electromagnetic
force is than the gravitational force, we divide the electromagnetic force by the gravitational force. The result is

$$\frac{\mathscr{E}}{\mathscr{G}} = \frac{K \times (\text{charge of proton}) \times (\text{charge of electron})/(0.53 \times 10^{-8} \text{ cm})^2}{-G \times (\text{mass of proton}) \times (\text{mass of electron})/(0.53 \times 10^{-8} \text{ cm})^2}$$

After canceling the $(0.53 \times 10^{-8}$ cm$)^2$ that appears the
same way in both the electromagnetic force and the
gravitational force, we have

$$\frac{\mathscr{E}}{\mathscr{G}} = \frac{K \times (\text{charge of proton}) \times (\text{charge of electron})}{-G \times (\text{mass of proton}) \times (\text{mass of electron})}$$

Notice that the part of the force that is the (distance
between centers)^2 *cancels* out because it appears exactly
the same way in both force laws.

We can numerically evaluate this ratio of electromagnetic force to gravitational force by using the numerical
values for K, G, the (mass of proton), (mass of electron),
(charge of proton), and (charge of electron). In terms of
these numbers[1] the ratio becomes

$$\frac{\mathscr{E}}{\mathscr{G}} = 2.2 \times 10^{39}$$

That is, the electromagnetic force between the proton
and electron in a hydrogen atom is more than 1,000 trillion trillion trillion (10^{39}) times larger than the gravitational force. This ratio is independent of the distance
between the proton and the electron because both forces
depend on the distance in the same way. The enormous
ratio shows that we can ignore the effects of gravitational forces inside atoms because electromagnetic forces
completely overshadow them.

Since the ratio of the electromagnetic force to the
gravitational force between two particles does not depend
on the distance between their centers, it is natural to
ask: "Why don't electromagnetic forces dominate gravitational forces at all distances?" "How can gravitational

---

[1] $K = 9 \times 10^{18} \dfrac{\text{gm} \times \text{cm}^3}{\text{coul}^2 \times \text{sec}^2}$ $\quad G = 6.67 \times 10^{-8} \dfrac{\text{cm}^3}{\text{gm}^2 \times \text{sec}^2}$

$M(\text{proton}) = 1.6724 \times 10^{-24}$ gm $\quad M(\text{electron}) = 9.1 \times 10^{-28}$ gm

$e = 1.6 \times 10^{-19}$ coul

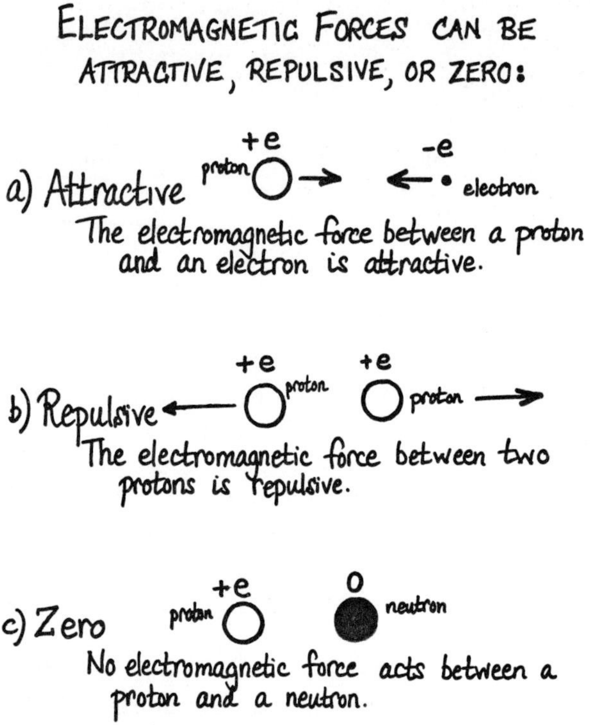

ELECTROMAGNETIC FORCES CAN BE
ATTRACTIVE, REPULSIVE, OR ZERO:

a) Attractive

The electromagnetic force between a proton
and an electron is attractive.

b) Repulsive

The electromagnetic force between two
protons is repulsive.

c) Zero

No electromagnetic force acts between a
proton and a neutron.

FIGURE 7
The electromagnetic force between two particles depends upon
the kind of electric charge that the particles have.

forces dominate electromagnetic forces at large distances?" To answer these questions and gain some in-
sight into Coulomb's law, we must look more closely a
examples of the electromagnetic force between two ele
mentary particles (Fig. 7).

First, we consider the electromagnetic force between a
proton and electron separated by a distance of 1 cm:

$$\mathscr{E} = \frac{K \times (+e) \times (-e)}{(1 \text{ cm})^2} = -\frac{K \times (e) \times (e)}{(1 \text{ cm})^2}$$

This electromagnetic force is a negative number (becaus
$1 \times -1 = -1$), and the force is attractive. (The sign con

vention for forces works properly if we just have no sign
in front of the K in Coulomb's law, but we need the
minus in front of the G in Newton's law to have things
come out right.)

Secondly, we consider the electromagnetic force between two electrons that are separated by the same
distance of 1 cm:

$$\mathscr{E} = \frac{K \times (-e) \times (-e)}{(1 \text{ cm})^2} = + \frac{K \times (e) \times (e)}{(1 \text{ cm})^2}$$

The product of the two negative charges is a positive
number, so the force is positive and repulsive.

Thirdly, and lastly, we consider the electromagnetic
force between a proton and a neutron that are separated
by 1 cm:

$$\mathscr{E} = \frac{K \times (+e) \times (0)}{(1 \text{ cm})^2} = 0$$

We find that neutral particles experience no electromagnetic forces because their electric charge is zero.

The last example will help us to explain why gravitational forces overpower electromagnetic forces for large
groups of particles at very large distances. At large distances, the atoms that form most objects appear to have
a net electric charge of zero. Consider two hydrogen
atoms, one on the moon and one on the earth (Fig. 8).
The distance between their centers is about 240,000
miles, the distance from the earth to the moon. The total
electric charge of a hydrogen atom is zero, and to another
hydrogen atom 240,000 miles away it simply appears to
be a single particle with no net electric charge. The
electromagnetic force between the two distant hydrogen
atoms is

$$\mathscr{E} = \frac{K \times (0) \times (0)}{(240,000 \text{ miles})^2} = 0$$

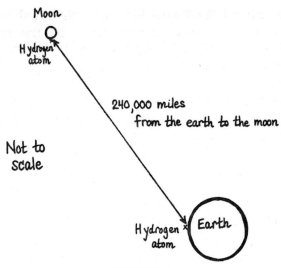

FIGURE 8

Two hydrogen atoms, one on the earth and one on the moon, are separated by 240,000 miles, the distance from the earth to the moon.

The electromagnetic force between these two distant hydrogen atoms (and between all atoms that are not close enough for the electrons and the nucleus to appear separate and distinct) is zero. The key to understanding this zero force is to realize that as long as the distances of separation are very large an atom appears to be a single neutral particle. This is not so for small distances, which is why electromagnetic forces bind atoms into molecules.

In contrast to the electromagnetic force, the attractive gravitational force between the two distant hydrogen atoms is not zero; it is

$$\mathscr{G} = -\frac{G \times (\text{mass of H atom}) \times (\text{mass of H atom})}{(240{,}000 \text{ miles})^2}$$

For any two atoms, this force is very small. However, there are about 10^{50} atoms in the earth and 10^{48} in the moon, and since the gravitational force between any two of these atoms is *always attractive* the total gravitational attraction between the earth and the moon is huge. This is a typical example of how gravity dominates when large distances and huge numbers of particles are involved.

The additive power of gravitational forces allows them to dominate other forces, once we consider large groups of particles like the earth and the moon. On the other hand, electromagnetic forces cancel out in large groups of particles because the attractive forces tend to cancel out in large groups of particles because the attractive forces tend to cancel the repulsive forces since there are about the same number of positive and negative electric charges. If a person's little finger had 1 percent more protons than electrons, that finger would produce an electromagnetic force of attraction on an electron on the moon that would be a trillion times stronger than the pull of the earth's gravity on the electron. But in fact, all large groupings of matter have an almost perfect balance of positive and negative electric charges, and so it is the always-additive gravitational forces that dominate large bodies separated by large distances, such as the earth and the moon.

VECTORS

A force like one of those described in this chapter can be characterized by two quantities: the *strength* of the force and the *direction* in which it acts. We can use an object called a *vector* to describe a force. A vector is an arrow; the arrow's length is proportional to the strength of the force it describes, and the arrow points in the direction of the force (Fig. 9). We shall use letters with arrows over them to indicate vectors. In Fig. 9, vector $\vec{B}$, which is twice as long as vector $\vec{A}$, describes a force twice as strong as vector $\vec{A}$.

We can use vectors to find the combined effect of two or more forces (Fig. 10). To add vectors $\vec{A}$ and $\vec{B}$ in Fig. 10

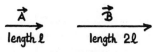

FIGURE 9
Two vectors that describe two forces acting from left to right.
Vector $\vec{B}$, twice as long as vector $\vec{A}$, describes a force twice
as strong as vector $\vec{A}$ does.

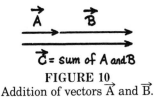

FIGURE 10
Addition of vectors $\vec{A}$ and $\vec{B}$.

we first place vector $\vec{B}$ at the head of vector $\vec{A}$. We then connect the base of vector $\vec{A}$ to the head of vector $\vec{B}$ and produce the sum of vectors $\vec{A}$ and $\vec{B}$: the new vector $\vec{C}$. In this particular example, vector $\vec{C}$ still points from left to right; $\vec{C}$ has a length of 3ℓ because $\vec{B}$ has a length of 2ℓ and $\vec{A}$ has a length of ℓ, and $\vec{A}$ and $\vec{B}$ point in the same direction. We also could have added vectors $\vec{A}$ and $\vec{B}$ by first placing the base of vector $\vec{A}$ at the head of vector $\vec{B}$ and then connecting the base of vector $\vec{B}$ to the head of vector $\vec{A}$ (Fig. 11), which would give us the same answer as before

We next consider a case where we add two forces (vectors) with the same strength but that point in opposite directions (Fig. 12). In this example, the sum of the vectors is zero because after the base of $\vec{B}$ is placed at the head of $\vec{A}$, the base of vector $\vec{A}$ and the head of vector $\vec{B}$ lie at the same point and produce a total sum of zero.

Figure 13 shows the addition of two vectors that are not parallel to one another. Once again we perform the addition by placing the base of $\vec{B}$ at the head of $\vec{A}$ and then connecting the base of $\vec{A}$ and the head of $\vec{B}$ to form the new vector $\vec{C}$.

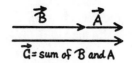

FIGURE 11
Addition of vectors $\vec{B}$ and $\vec{A}$.

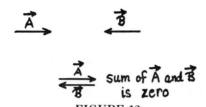

FIGURE 12
Addition of vectors pointing in opposite direction.

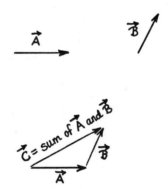

FIGURE 13
Addition of two vectors not parallel to one another.

SUMMARY

There are four types of forces in the universe: gravitational, electromagnetic, strong, and weak. Gravitational forces are always attractive. Their additive power allows them to dominate all other forces for large groups of particles, like the sun and the earth, separated by large distances. Electromagnetic forces attract particles with opposite signs of electric charge and repel particles with the same sign of electric charge. They dominate other forces over distances the size of atoms and can bind atoms together into large chains of molecules. Strong forces hold the nuclei of atoms together but fall off rapidly for distances greater than nuclear sizes. They dominate over distances about the size of a proton's radius. Weak forces operate at the same distances (or less) as strong forces. They are responsible for some of the details of elementary particle interactions and are a necessary part of the chain of reactions that liberate energy in stars. A typical weak reaction is the decay of a neutron into a proton, an electron, and an antineutrino. Whenever neutrinos or antineutrinos are involved, we have a weak reaction.

QUESTIONS

1 Do gravitational or electromagnetic forces determine the motion of the planet Mars around the sun?
2 In the nucleus of a typical nitrogen atom there are seven protons and seven neutrons. Do the protons repel the neutrons electromagnetically? What force overcomes the protons' electromagnetic repulsion of one another?
3 A water molecule is formed when two hydrogen atoms attach themselves to an oxygen atom. What kind of force produces this attachment?
4 Do strong forces attract and repel, or do they only attract?
5 Recall our scale model for atoms and their nuclei. Suppose that protons were the size of grapefruit and that two protons (grapefruit) were separated by 10 miles. What force would dominate this situation?
6 Suppose now that the two protons (grapefruit) were separated by 10 in. Which force would dominate this situation in our scale model?

7 For a proton and an electron separated by 3 cm, calculate how many times larger the electromagnetic force between them is than the gravitational force between them. (*Ans.*: 2.2×10^{39})

8 Consider the following system of particles: From left to right along a straight line we have a proton, an electron, and a helium nucleus. The distance between the proton and the electron is 5 cm, and the distance between the electron and the helium nucleus is 10 cm.

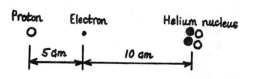

(*a*) In what direction does the electromagnetic force from the proton act on the electron?

(*b*) In what direction does the electromagnetic force from the helium nucleus act on the electron?

9 If a set of forces acts in a line on a particle, then the total force acting on the particle is the sum of these forces, if proper account is taken of the direction of these forces. In Prob. 8, in what direction does the total force from the proton and the helium nucleus acting on the electron point? Which of the following gives the size of the force?

(*a*) $\dfrac{K \times (e) \times (e)}{(5 \text{ cm})^2}$ (*b*) $\dfrac{K \times (e) \times (e)}{(10 \text{ cm})^2}$

(*c*) $\dfrac{1}{4} \dfrac{K \times (e) \times (e)}{(5 \text{ cm})^2}$ (*d*) $\dfrac{1}{2} \dfrac{K \times (e) \times (e)}{(5 \text{ cm})^2}$

10 If a negative pi meson (π^-) were to decay into an electron (e^-) and an antineutrino ($\bar{\nu}_e$), what sort of a reaction would occur?

5
matter
and
antimatter

*The large steel door clanged behind Zed and his
assistant, and Cyril Zaki heard their footfalls die into
a metallic haze. An icy excitement gripped Zaki's bowels.
The minutes were so short and his need so great.
Quickly he lit a melloroon and raised himself to a
sitting position.*

*His piercing grey eyes scanned the sides of his prison.
Darting, probing, they noticed almost everything: the
viewtapes with their strange titles, the computo-consoles,
the strange, glittering equipment. Beads of sweat
appeared on Zaki's forehead as he stood. Gliding toward
the wall with the viewtapes, he quickly pulled out the
tape marked Candy and carefully placed it out of harm's
way. Years as a top insurance agent had taught him to
respect property whenever possible.*

*"Hi curootie," Cyril bellowed as he set himself into
a number six position with feet spread apart and firmly
planted.*

*"Zungi!" Cyril's left hand darted into the vacated
cubbyhole and smashed for freedom.*

*"EEyah!" cried Cyril as he withdrew his aching
hand with the realization that the rear of the cubbyhole
was sturdier than he had hoped. None of that pasteboard
stuff like the Trust's library on Pilar, he mused dolefully*

as he rubbed his sore hand. Using his still functioning right hand, Cyril lifted his melloroon and took a long toke.

"Cough, gasp, damn that Zed giving me stale zingers."

Zaki threw down the half-smoked melloroon and crossed to the part of the room with the most unfamiliar equipment. Some instinct, perhaps that same sure grasp of the essential that had carried him to such heights in the Trust, told him that the clue to escape was here. Cyril was a man in touch with himself and knew that in solving problems his hunches had sometimes even more value than his capacity for reason.

Zaki examined three huge doughnut-shaped blocks made of a strange shiny material, unlike even the purium used in his spaceship. Clear plastic tubes carrying vari-colored fluids pierced the blocks at odd angles. Strips of red metal surrounded the glistening doughnuts, isolating them from the rest of the room. Conduits of wire issued from the blocks like huge unruly masses of black hair, leading to a control panel directly to Cyril's right hand.

Sweat gathered in Zaki's armpits. He knew that the answer was right here if he could just puzzle it out. He began to examine the two rows of touch-bars. On the top left, the green bar glowed its title PROTON BEAM. To its right the other touch-bar stated ANTIPROTON BEAM. Below these two were two blue touch-bars that read ELECTRON BEAM and POSITRON BEAM. But Cyril's attention flashed on the two bars winking on and off in red, proclaiming MATTER and ANTIMATTER. Deep in his zor, he knew the answer must be in front of him. Sweat poured down Cyril's shift as he stood before the control panel. If antimatter is anti to matter, he thought, then matter and antimatter must somehow be opposite, so if we play one off against the other . . . . With trembling hands he reached forward and pushed the two flashing red bars MATTER and ANTIMATTER. Zaki's stomach was in horrible knots.

A hissing sound swelled and filled the dome-shaped room. Cyril backed away from the control console as the

static changed into a loud humming. Suddenly the room
vibrated with, "I'll sign your policy, baby . . . ," the
plaintive wail of Johnny Jumbo's smash hit, now ten
accounting periods old.

"Great cancellation," muttered Zaki, as he lurched
toward the console. He had hated this melodoo even
when it was at the top of the charts. In a frenzy of anger,
Zaki's huge hands reached out and pressed the PROTON
and ANTIPROTON bars.

Johnny Jumbo was canceled abruptly, replaced by
silence. Relief flooded through Zaki like a water closet in
action. But almost at once distant footsteps announced
the impending return of Dr. Bokomoru Zed.

In a panic, Cyril started to run from one part of the
room to another, probing and thrusting roughly for any
avenue of escape. Because of the tears in his eyes, he was
unable to see anything properly.

Thump. "Aagawa!" Zaki's huge thigh collided with
the steel disposal bin, ripping it right off the side of the
wall. Cyril turned quickly to examine the damage. His
instincts for a possible premium were always good. In
the side of the wall, where the steel bin had hung, a
three-foot wide shiny disposal tube gaped an invitation.
Cyril stood rubbing his eyes while childhood thoughts
of claustrophobia and huge worms welled up in him.
The worms got my sister, he thought. I'll never see
Vibeke again.

Click. Zed was opening the massive door.

Zaki made a sudden decision. It was either trash
himself into the tube, or more of Zed. An easy choice,
Cyril thought, as he stuck his head into the tube.

Into the just-vacated room came Bokomoru Zed and
his assistant. Bokomoru looked slightly disheveled and
eminently pleased with himself. "I thought he'd never
escape." laughed the blond assistant with the perfect zor.

"I was beginning to have some doubts myself," replied
Zed. "We must find out whether Borg knows where the
Permids are. I knew that Borg would send an investigator

once we were established near the black hole, but I never thought she'd use an incompetent like Zaki."

"Well, Boko, Zaki must be her best man. He's got the biggest premium in the Trust."

"Just a poor excuse for a spy. But with the sender inside him, we can make him our own probe and get the goodies direct from Pilar," Bokomoru concluded with a slurp as he patted his assistant's zor and peered down the tube that Cyril had just traversed.

At the far end of that tube Cyril Zaki was emerging into the warm air of Ripov. He turned to examine the grey building from which he had escaped. "Gasp," he choked; it was one of the museums that Lisa Dalby had pointed out to him. As usual, Cyril lost no time in reflection on strange coincidences before taking action. He had to round up his crew and report back to Borg as soon as he could.

What is antimatter? And what would happen if matter and antimatter were to come together?

Ordinary matter is composed of elementary particles like protons, neutrons, and electrons, the constituents of atoms. But for every kind of elementary particle there is a corresponding kind of antiparticle. Antimatter consists of these antiparticles. An antiuniverse would be made up of antiatoms. Each antiatom would have a tiny nucleus made of antiprotons and antineutrons, orbited by distant antielectrons (positrons). Before we consider what happens when matter and antimatter meet, we shall describe some particles and their antiparticles.

All elementary particles and their associated antiparticles can be divided into a class structure. Each class is distinguished from the others by the functions its member particles play in preserving the order of the universe as we know it.

An atom contains representatives of two classes of particles. The electrons in orbit around the nucleus are members of the class called *leptons*, and the protons and neutrons that form the nucleus are members of the class called *hadrons*. In this chapter we discuss leptons and antileptons, hadrons and antihadrons, and then we bring matter and antimatter together.

The chief representative of the class of leptons is the electron. In Table 1 we describe two of the most common leptons: electrons and neutrinos. A neutrino has no mass and no electric charge. A particle with no mass can never come to rest. Instead, it always moves at the speed of light.

An antiparticle has the same mass as its corresponding particle, but the antiparticle has the opposite sign of

TABLE 1

Particle name	Symbol	Mass, gm	Electric charge
Electron	e^-	0.91×10^{-27}	$-e$
Neutrino	ν_e	0	0

TABLE 2

Particle name	Symbol	Mass, gm	Electric charge
Antielectron or positron	e^+	0.91×10^{-27}	$+e$
Antineutrino	$\bar{v}_e$	0	0

electric charge. In Table 2 we describe the two anti-leptons that correspond to the two leptons of Table 1.

The antielectron or positron has the same mass as the electron, but it has one unit of positive instead of negative charge. The antineutrino, like the neutrino, has no mass and no electric charge, since zero with the opposite sign is still zero. However, a neutrino and an antineutrino interact with other elementary particles in different ways, and this difference allows us to classify them as different kinds of particles. For example, when a neutrino (v_e) collides with a neutron, afterward a proton and an electron can appear (Fig. 1). But when an antineutrino ($\bar{v}_e$) hits a neutron, the neutron and antineutrino remain unchanged (Fig. 2).

What necessary functions do the leptons and antileptons fulfill in the preservation of order in our universe? There are two principal ones. First, the electrons are constituents of the atoms from which all matter is formed. Secondly, the leptons and antileptons can participate in "weak forces" or "weak interactions." Whenever neutrinos or antineutrinos appear, we get a weak inter-

FIGURE 1

Before **After**

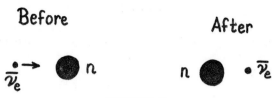

FIGURE 2

action that involves weak forces. In stars like the sun, these weak interactions are a major part of the reactions that convert some of the sun's energy of mass into energy of motion (kinetic energy). This energy of motion is eventually transformed into the light and heat that reaches us on earth.

Among elementary particle classes, leptons differ from more massive particles like protons and neutrons because leptons *do not feel strong forces.* Particles that experience strong forces are called *hadrons.* Hadrons are perhaps the most important, and as yet the most mysterious, of the elementary particles. In general, hadrons are much more massive than leptons. The functions of the hadrons and their antiparticles are also twofold. First, the best-known hadrons, protons and neutrons, form the nuclei of atoms and hence comprise most of what we call matter. Secondly, the hadrons themselves produce the "strong" forces that bind the nuclei of atoms. Table 3 lists some of the most common hadrons and their corresponding antihadrons. An antihadron has the same mass as its corresponding hadron, but it has the opposite sign of electric charge.

There are more hadrons than those described in Table 3. In fact, there may be an infinite number of types of hadrons with greater and greater masses. At present we are beginning to arrange the different species of hadrons in a crude periodic table, much as the different species of atoms were cataloged in a periodic table 100 years ago.

The only really stable particles are protons, electrons, neutrinos, photons, and their corresponding antiparticles. All other elementary particles eventually decay into

TABLE 3

Hadrons

Name	Symbol	Mass, gm	Electric charge
Proton	p	1.6724×10^{-24}	$+e$
Antiproton	$\bar{p}$	1.6724×10^{-24}	$-e$
Neutron	n	1.6747×10^{-24}	0
Antineutron	$\bar{n}$	1.6747×10^{-24}	0
Pi-plus meson	π^+	0.2488×10^{-24}	$+e$
Anti-pi-plus or pi-minus	π^-	0.2488×10^{-24}	$-e$
Pi-zero meson	π^0	0.2405×10^{-24}	0
Anti-pi-zero or pi-zero	π^0	0.2405×10^{-24}	0
Pi-minus meson	π^-	0.2488×10^{-24}	$-e$
Anti-pi-minus or pi-plus	π^+	0.2488×10^{-24}	$+e$

these four types, provided they are not near other elementary particles with which they can interact.[1]

When a particle and its antiparticle meet, they can completely annihilate one another, turning *all* their energy of mass into energy of motion. For example, consider an electron and positron that are about to collide (Fig. 3). The electron and positron can completely transform all their energy of mass into two particles with no mass called photons (Fig. 4). In this process, all the energy of mass is converted into the energy of motion of the photons. Photons are the elementary particles that form light; they always travel at the velocity of light (186,000 miles/sec). Photons and antiphotons are identical, so they are both called photons.

In the sun, the positrons produced in the first step of the proton-proton cycle (see Fig. 6, Chap. 3) annihilate

---

[1]There has been some recent speculation that protons, neutrons, and indeed all the hadrons are made up of superelementary particles called *partons* (part of a hadron). These partons may orbit one another in a manner similar to the electron orbiting the proton in a hydrogen atom. However, there is no definite experimental evidence that partons exist.

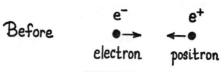

FIGURE 3

An electron and a positron that are about to collide.

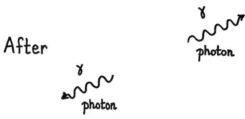

FIGURE 4

Two photons are produced after the electron and positron annihilate.
They leave the scene of annihilation in opposite directions.

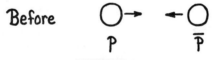

FIGURE 5

An almost stationary proton and antiproton near one another.

with the electrons they meet in the sun's interior. This mutual annihilation adds to the energy of motion that is liberated by nuclear reactions inside the sun and in other stars.

What happens when a proton meets an antiproton (Fig. 5)? One possible outcome of this meeting is the complete annihilation of the proton and the antiproton into four photons (Fig. 6). After the collision, all the energy of mass of the proton and antiproton has been converted into the energy of motion of the four massless photons, each traveling with the speed of light.

Another possible outcome of a proton and antiproton coming together is their annihilation into a neutrino,

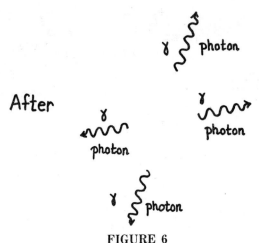

FIGURE 6
The annihilation can produce four photons.

antineutrino, and six photons (Fig. 7). In this process, all the energy of mass is converted into the energy of motion of the neutrino, antineutrino, and six photons.

When matter (particles) and antimatter (antiparticles) are brought together, there is always the possibility that all their mass will be annihilated and transformed into the energy of motion.[1] Usually this energy of motion will be in the form of photons, neutrinos, and antineutrinos, all of which are massless particles traveling with the speed of light. In contrast, when ordinary matter (particles) collides with ordinary matter (particles), only a *small fraction* of the mass involved can be converted into energy of motion. Each kind of particle has a specific kind of antiparticle with which it can perform a total mutual annihilation. If a woman made of matter and a man made of antimatter were to come together, they would get off in a blaze of energy greater than 100,000 hydrogen bombs.

Our solar system, and in particular, the earth, is composed almost entirely of ordinary matter. If it were half

---

[1] In the proton-proton cycle discussed in Chap. 3, only about 1 percent of the energy of mass of the original particles is converted into energy of motion.

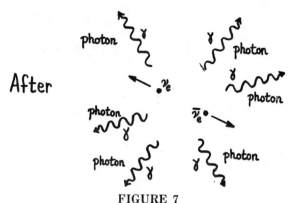

FIGURE 7
The annihilation can produce six photons,
a neutrino, and an antineutrino.

matter and half antimatter, it would very quickly annihilate itself into pure energy of motion, with no earth
left under our feet. It *is* possible that half the universe
could be made of matter (particles) and the other half of
antimatter (antiparticles). In the antimatter parts of the
universe, antiatoms would be composed of antiprotons,
antineutrons, and antielectrons (positrons). The antiatoms would also be held together by electromagnetic
forces, just as ordinary atoms are, but the antiatoms
would have a negatively charged nucleus orbited by positively charged positrons. The antimatter parts of the
universe would have to be separated from our "ordinary"
matter, or mutual annihilation would occur. Nonetheless, it is possible that a small antimeteor from the antimatter of the universe could eventually stray into our
matter part of the universe and create a huge explosion
when it finally collided with the ordinary matter of a
planet like our earth. But so far we can not tell just how
much of the universe around us is made of antimatter
rather than ordinary matter because photons that come
from stars are the same as those which would come from
antistars.

At present the only antiparticles on earth are those
that are created in huge particle accelerators, and they

Before

FIGURE 8
Two slowly moving protons about to collide.

do not exist long because they quickly annihilate when they collide with ordinary particles around them.

To understand how strong and weak forces appear when elementary particles interact, we shall consider the interior of our sun. In the sun, the most important series of energy-producing reactions among particles is the proton-proton cycle, which consists of three steps. We shall describe the first step of the proton-proton cycle as an example of a combination of both strong and weak interactions. This step also illustrates how some of a star's energy of mass is converted into energy of motion.

The proton-proton cycle begins when two slowly moving protons collide. These protons are moving so slowly that we may neglect their energy of motion (Fig. 8); they have just enough energy of motion so that they can bump into each other.[1]

After the collision, the two protons have been transformed by a combination of strong and weak interactions into a deuteron, positron, neutrino, and some additional energy of motion (Fig. 9). The deuteron has one unit of positive charge ($+e$) and a mass of 3.3432×10^{24} gm.

The production of the deuteron, positron, and neutrino takes place in two steps: a strong interaction and a weak interaction. We shall describe one possible sequence of events in these reactions (Fig. 10).

---

[1] By our standards on the earth, the protons' speed is not so small, namely several hundred miles per second. However, the two protons' initial kinetic energy is several hundred times smaller than the final kinetic energy that is liberated by the conversion of some of the energy of mass of the initial two protons into energy of motion. The initial motion of the two protons is just adequate for them to overcome their own mutual electromagnetic repulsion and bump into one another.

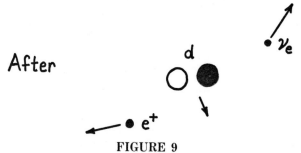

11

After

FIGURE 9
After the collision, we may find a deuteron (d), a positron (e^+),
an antineutrino $(\bar{v}_e)$, and some additional energy of motion.

First, the two protons come near each other, and one of the protons transforms into a neutron, positron, and neutrino. This is a weak reaction because a neutrino appears.

Secondly, the proton and the (transformed) neutron, being close to one another, are bound into a deuteron by strong forces. The net result of these two steps is the production of a deuteron, positron, and neutrino as well as some additional energy of motion.[1]

Energy of motion is liberated in this combination of strong and weak reactions because the combined mass of the two protons before the collision is greater than the combined mass of the deuteron, positron, and neutrino after the collision (see Table 2, Chap. 3, page 72). This means that some of the mass of the two protons was converted into the energy of motion (kinetic energy) of the deuteron, positron, and neutrino. Usually the positron and neutrino carry off most of this kinetic energy; only about 0.02 percent of the two protons' energy of mass is converted into the energy of motion. Through a series of reactions similar to the one just described, part

---

[1] These two steps can take place in either order. For example, the two protons could first be bound by strong forces. Then one of the protons could transform via a weak reaction into a neutron, positron, and neutrino. This would also leave a proton and a neutron bound by strong forces to form a deuteron as well as a free positron and neutrino.

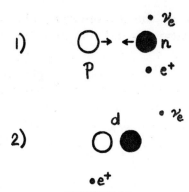

FIGURE 10

The process may occur when one of the protons transforms
into a neutron, positron, and antineutrino. The proton
and neutron then combine to form a deuteron.

of the sun's energy of mass is gradually being converted
into kinetic energy, which finally reaches us in the form
of light and heat.

We now consider an important conservation law: the
conservation of total electric charge. In any reaction
between elementary particles and/or antiparticles, the
total electric charge before the reaction is the *same* as
the total electric charge after the reaction. To illustrate
this concept, we consider two reactions. First we consider
the initial step of the proton-proton cycle, where two protons collide to produce a deuteron, positron, neutrino,
and some energy of motion (Figs. 8 and 9).

Before	*After*
Symbol:	Symbol:
p Electric charge of a proton $= +e$	d Electric charge of a deuteron $= +e$
p Electric charge of a proton $= +e$	e^+ Electric charge of a positron $= +e$
	v_e Electric charge of a neutrino $= 0$
Total electric charge before $= +2e$	Total electric charge after $= +2e$

Total electric charge before = Total electric charge after

Before the collision, the two protons had a total electric charge of $+2e$. After the collision, the deuteron (electric charge $+e$), positron (electric charge $+e$), and neutrino (electric charge 0) have a total charge of $+2e$. Even though particles may be transformed into other particles, the *total electric charge always remains the same*.

Secondly, we consider the reaction where an electron and a positron annihilate into two photons (Figs. 3 and 4).

Before	*After*
Symbol:	Symbol:
e^- Electric charge of an electron $= -e$	γ Electric charge of a photon $= 0$
e^+ Electric charge of a positron $= +e$	γ Electric charge of a photon $= 0$
Total electric charge before $= 0$	Total electric charge after $= 0$

Total electric charge before = Total electric charge after

The total electric charge of the electron (charge $-e$) and positron (charge $+e$) system before the annihilation is zero, and afterward the total electric charge of the two photons is zero. In any reaction total electric charge is always conserved.

Whenever one hadron comes close to another hadron, the two hadrons can interact through strong forces (also called strong interactions). ("Close" means distances of 5×10^{-13} cm or less.) Leptons, such as electrons and positrons, do not participate in the strong forces. These strong forces or strong interactions can occur when two rapidly moving hadrons collide or when two hadrons are located near one another, like the protons and neutrons in an atom's nucleus.

We now attempt to explain how the strong forces arise and the role of the hadrons themselves in producing these forces. To do this we shall describe why a neutron does not decay in 1,000 sec into a proton, electron, and antineutrino when the neutron is in a nucleus and near other hadrons. Consider a deuteron, which is the nucleus of a deuterium atom (an isotope of hydrogen). The deuteron consists of one proton and one neutron bound by strong forces (Fig. 11).

We shall describe one possible sequence of events that intervenes to prevent the decay of the deuteron's neutron in 1,000 sec into a proton, an electron, and an antineutrino. This sequence takes place in an incredibly short period of time, about 30 trillion-trillionths (30×10^{-24}) of a second, which is many times less than the time needed for an isolated neutron to decay.

THE SEQUENCE

1   At time $t = 0$ sec, the proton and neutron are separated by about 3×10^{-13} cm (Fig. 12).

2   After a time $t = 10$ trillion-trillionths of a second (10^{-23} sec), the proton transforms into a stationary neutron and a pi-plus meson (π^+) headed toward the neutron. The pi-plus meson is moving at almost the speed of light (Fig. 13).

3   After another 10 trillion-trillionths of a second (10^{-23} sec), the pi-plus meson reaches the right-hand neutron (Fig. 14).

FIGURE 11
A proton and a neutron form the nucleus of a deuterium atom.
The nucleus of a deuterium atom is also called a deuteron.

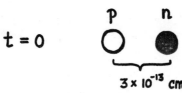

$t = 0$

P   n

3×10^{-13} cm

FIGURE 12
Proton and neutron separated by 3×10^{-13} cm.

$t = 1 \times 10^{-23}$ sec

n   n

π^+

FIGURE 13
The proton temporarily becomes a neutron and a pi-plus meson (π^+).
The pi-plus meson is headed toward the right-hand neutron.

$t = 2 \times 10^{-23}$ sec

n   n

π^+

FIGURE 14
The pi-plus meson (π^+) reaches the right-hand neutron.

4   After another 10 trillion-trillionths of a second (10^{-23} sec), the right-hand neutron absorbs the pi-plus meson and is transformed into a proton (Fig. 15).

5   Now, after a total elapsed time of 30 trillion-trillionths of a second (3×10^{-23} sec), the right-hand neutron has become a proton long before it could decay (in 1,000 sec) into a proton, an electron, and an anti-neutrino (Fig. 16). The original left-hand proton is now a neutron, so the whole cycle can reverse direction and start over.

We have described just one possible sequence of events, in which a proton and a neutron exchange a pi-plus meson, changing the proton into a neutron and the neutron into a proton. This exchange of mesons actually prevents the neutron from decaying and helps bind the

FIGURE 15
The pi-plus (π^+) meson is absorbed by the neutron, and
the neutron becomes transformed into a proton.

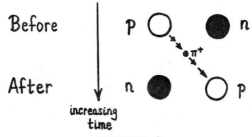

FIGURE 16
The proton is now a neutron and the neutron is now a proton, and
so the whole cycle can start again in the opposite direction.

Before

After

increasing
time

FIGURE 17
A schematic version of the time sequence of the exchange
of a pi-plus meson (π^+) between a proton and a neutron.

neutron and the proton. Figure 17 is a combination of
Figs. 12 to 16 and shows a schematic version of the
exchange of a pi-plus meson between a proton and a
neutron.

Another possible sequence events, which is similar but
not identical to the previous one, is the exchange of a
pi-minus meson (π^-) between a neutron and a proton,
represented in Fig. 18. The neutron starts things by
decaying into a proton and a pi-minus meson, and the
proton subsequently absorbs the pi-minus meson.

Another set of events involves the exchange of a
neutral pi meson between the proton (which remains a

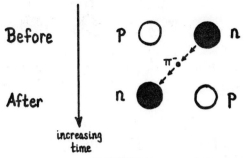

FIGURE 18

A schematic version of the time sequence of the exchange of
a pi-minus meson between a neutron and a proton.

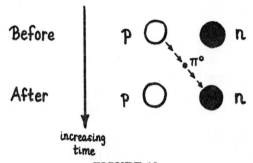

FIGURE 19

A schematic version of the time sequence of the exchange of
a neutral pi meson (π^0) between a proton and a neutron.

proton) and the neutron (which remains a neutron). This
is represented in Fig. 19.

All the processes described above take place over and
over when two hadrons are close to one another. We can
not observe the pi mesons that are exchanged between
the protons and the neutrons, and so such pi mesons are
called "virtual" rather than real particles.

We believe that all forces between elementary particles
are transmitted by the exchange of other elementary
particles in the virtual state of existence. For example,
electromagnetic forces are transmitted between particles
with electric charge by virtual photons (Fig. 20).

122

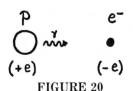

FIGURE 20
Electromagnetic forces are transmitted by virtual photons.

The virtual particles that transmit forces travel with the speed of light (180,000 miles/sec) so that the effect of the different types of forces travels with the speed of light. Since these virtual particles are invisible, forces appear to act by invisible wires, but actually they are transmitted by elementary particles traveling at the speed of light.

Each type of force is transmitted by the exchange of a particular virtual particle or group of virtual particles. Usually the real form of these virtual particles has been observed directly. We can now list three of the four types of forces and the particles that transmit them:

1   Strong forces are transmitted mainly by virtual pi mesons. To a lesser degree, all the other hadrons in their virtual-particle forms help transmit the strong forces. Pi mesons and other hadrons have been detected in their real form.

2   Electromagnetic forces are transmitted by virtual photons. Real photons in the form of light, radio waves, and so forth have been detected many times.

3   Gravitational forces are transmitted by particles called *gravitons*. Until now, no one has successfully detected a graviton.

SUMMARY

Elementary particles can be grouped into two classes: leptons (electrons and neutrinos) and the much heavier

hadrons (for example, protons, neutrons, and pi mesons).
Only the hadrons experience strong forces. Each variety
of elementary particle has a corresponding antiparticle
variety. The antiparticle has the same mass but the
opposite sign of electric charge. If a particle meets its
antiparticle, the two annihilate completely. They turn
all their energy of mass into the energy of motion of
photons, neutrinos, and antineutrinos. Photons form a
separate class of particles, and a photon and antiphoton
are identical. In *any* process, such as the annihilation of
a particle and its antiparticle, the *total* electric charge
stays the same.

QUESTIONS

1 Why are hadrons different from other elementary
  particles?
2 In the first step of the proton-proton cycle (p. 114), a
  positron is produced. What could happen inside the
  sun if such a positron met an electron?
3 In the proton-proton cycle, about 4 percent of the
  mass of a proton is converted into kinetic energy. If
  instead of taking place with nuclear reactions among
  ordinary particles, the proton-proton cycle consisted
  of a proton meeting an antiproton, what fraction of
  the energy of mass could then be transformed into
  the energy of motion?
4 What electric charge does an antiproton have? What
  electric charge does an antineutron have?
5 How many times more kinetic energy is liberated
  when a proton and an antiproton annihilate into
  two photons than when an electron and a positron
  annihilate into two photons? (*Ans.*: 1,836 times)
6 Consider a high-speed collision between two protons. After the collision, we find a proton, a neutron,
  a neutral pi meson (π^0), and one other pi meson.
  What electric charge does the other pi meson have?

7 Consider another high-speed collision between two protons, where a proton, a neutron, and *three* pi mesons are produced. What charges can the pi mesons have?

8 Calculate, in ergs, the amount of kinetic energy that would be released if 1 gm of matter and 1 gm of antimatter were to annihilate completely. (For comparison, human beings on earth use about 9×10^{19} ergs of energy each day.) (*Ans.:* 9×10^{20} ergs)

9 Suppose that the proton-proton cycle took place in antimatter as an antimatter cycle. In the first step of this cycle, two antiprotons collide. What particles would be the result in the antiproton-antiproton cycle?

10 Just as positrons and electrons can annihilate to produce two photons, two photons can sometimes turn into an electron-positron pair. Suppose that one electron and one positron are produced from two photons. What is the minimum energy of motion that the original photons must have had?

11 What kinds of particles transmit strong forces?

12 What kinds of particles transmit electromagnetic forces?

13 What type of forces or interactions take place during the first step of the proton-proton cycle?

6
time
dilation

Cyril Zaki floated in his dream world down a flowered staircase that became a tropical forest as he tried to reach the bottom. From a ferny undergrowth, gentle animals hummed space songs and then became a hundred flying thoughts that smoothed the corners of his mind as they winged by. Waves of startling pleasure rocked back and forth, and Zaki found himself before a mighty crowd, wearing a soft and shiny tunic to receive the homage due him. The people raised him on their hands, then bounced him with ever-growing shouts on a super-trampoline, higher and still higher. As Zaki soared toward the opening sky, he met an all-wise woman wearing a filmy shift. Her hair was like red cornsilk, and she gathered Cyril toward her soul as they sailed farther and farther, onward and deeper, until he went through the starshine and came into a bright and airy void.

"Welcome home, Captain," said the attendant as she sponged Zaki down. "Have another good trip?"

You couldn't help admiring Zaki. The man was certainly tough to let his mind and body both be twisted through the vise of special relativity.

"Aagh, great. AA. Quite secure," Cyril mumbled. Pleasant as these psychorama deceleration trips were, there was no avoiding that feeling of loss at the end.

Still, better that than losing your marbles like the first people to undergo massive time dilation. Human bodies were just not made for that kind of shock, but the high premium on space pilots made stretching lives worth the risk and difficulty. A good pilot would last for a long time — as measured at the terminal.

"When is it?" Cyril asked.

"You've been gone 9.4 years, terminal time."

To Cyril and his crew, the expedition had lasted about ten weeks. Zaki congratulated himself silently. Got through another, he thought. Time for the standard question.

"What's the good news here?"

"Well, the Jolters took the title in '14, the Bruisers in '15, the Bombers in '16, the Crushers in '17 and '18, the Bruisers again in '19, the Stompers in '20, the Rovers in '21, and the Uptights won this year."

"Anything else?"

"Nothing much, Captain, everything's been secure here. Your health code checks out AA. The car's waiting outside to take you to the I.E. Peace and security."

Zaki returned the admiration with a hero's insouciance.

"Your security as well."

He stepped into the aircar and soared away over Pilar to report to Zenith Borg.

Cyril glanced out the side viewport at the Pilar skyline. Rank upon rank of tall grey buildings lined the narrow concrete walkways like sentinels guarding the passage of a potentate. Zaki sensed that in some strange way it was he and his existence that these gray soldiers protected. Only a small fraction of the buildings were occupied; the rest held treasured equipment not currently operable. Below him, Cyril saw a few pedestrians threading their way among the buildings, though most of the traffic consisted of aircars like his own. A handful of people dared to use the barely functioning under-ground slot system, which once had floated most of the planet's commerce on magnetic fields in deep tunnels. From the outskirts of Pilar, a brown plane of haze stretched to the horizon, giving way in spots to the remaining green farmlands. In the gaps between the outer ranks of buildings, small domes enclosed the nuclear-powered air filters that struggled ceaselessly to keep the atmosphere breathable. Zaki reflected on what he saw. Dangerous things, those slots, he decided. Better to be out in the open where you can look around. He hadn't remembered Pilar as looking quite this ugly. Did Zed have a point after all? Cyril turned his gaze inward and contemplated his premium.

When Cyril Zaki takes a hit off his melloroon, his subjective sense of time changes. He becomes more mellow, more relaxed, a man in tune with his essence. But one man's inner time is often hard to share with another's. So Cyril in his generosity turns on the whole crew to time dilation. As the Top Dog approaches the speed of light, real time slows down for Cyril and his happy followers. They all share in the dilation of time. How can we on earth share in this experience? The easiest way to expand our understanding of how time slows down is through the contemplation of the hadrons, especially the neutron.

The decay of particles like the neutron can be used to verify the *relative nature of time*, first elucidated in Einstein's theory of special relativity. This relativity theory shows that for a system moving at high speeds with respect to us, time passes *more slowly* in that system than time within our system at rest. The faster the system travels, the more slowly we observe time to pass within that system. We can get our first glimpse of this phenomenon by comparing the decay of a neutron while it is moving with the decay when the neutron is at rest. Consider a motionless neutron, isolated from other particles.

It has a lifetime of 1,000 sec, after which we observe it decay into a proton, an electron, and an antineutrino (Fig. 1). If we now observe another neutron moving at nine-tenths the speed of light, we find that it takes 2,300 sec for this neutron to decay (Fig. 2). This observation tells us that in some way time slows down by a factor of

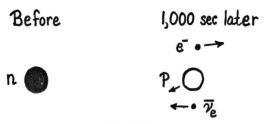

FIGURE 1
A motionless neutron decays after 1,000 sec.

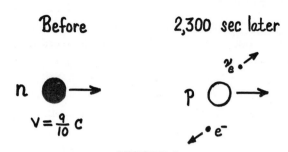

FIGURE 2
A neutron traveling at nine-tenths the speed of light
decays in 2,300 sec.

2.3 for a neutron when it moves with nine-tenths the speed of light.

At rest, the neutron's internal alarm clock tells the neutron to decay after 1,000 sec. If the neutron is moving at nine-tenths the speed of light, its clock appears to slow down by 2.3 times since the neutron decays after 2,300 sec. If a neutron travels at ninety-nine-hundredths the speed of light, its clock appears to slow down by 16.3 times, and we observe it to take 16,300 sec to decay.[1] Experiments have shown that the faster a system travels, the more slowly time passes within it.[2] This effect is known as *time dilation*, and it occurs for groups of particles as well as for individual elementary particles. Time dilation helps us to study elementary particles in laboratories because the particles live longer before they decay if they are accelerated to high velocities in giant particle accelerators.

---

[1] In its own reference system, the neutron always takes (or thinks it takes) 1,000 sec to decay, but to an outside observer this decay *appears* to take longer if the neutron is moving at high speed. Rapid motion causes the neutron's internal clock to slow down with respect to an external, stationary clock.

[2] Mathematically, the relationship between time in a system in motion with speed v and time in a system of stationary particles is

$$\text{Time in stationary system} = \frac{\text{time in moving system}}{\sqrt{1 - (v^2/c^2)}}$$

Relativity theory, which predicts the time dilation effect, has been verified thousands of times at the elementary particle level: particles moving at high speeds in accelerators take longer to decay into other kinds of elementary particles than if they are almost motionless. In addition, in a recent experiment a tremendously accurate atomic clock was carried around the world on a jet plane and compared with a similar clock that did not make the trip. Even though the speed of the jet plane was far, far less than the speed of light, the clock aboard the plane ran a trifle slower than the clock on the ground, and this difference could be measured. Thus it is correct to say that airplane pilots and astronauts are a fraction of a second younger than they would be if they had stayed at home all their lives.

Time dilation could someday have profound consequences for space pilots and crews. If spaceships were to travel at speeds near the speed of light, time would slow dramatically for the people aboard. What appeared to them to be a 1-year trip could well be a 20-year one by earth standards. In some cases, space crews could return to earth even younger than their children or grandchildren. This slowing of time in a fast-moving spaceship might eventually allow people to reach a star like Arcturus, about 35 light years away. (Light from Arcturus, traveling at 186,000 miles/sec, takes 35 years to reach us.) However, by the time these travelers returned to earth all the people who sent them on their journey might have died.

We now consider the relative nature of motion and discuss the famous "twin paradox." Consider two twins, a man and a woman: the man stays on earth, and the woman sails into space at nine-tenths the speed of light. The twin on earth sees his sister traveling at nine-tenths the speed of light and concludes that she is aging less rapidly than he is. On the other hand, the twin in the spaceship can insist that the earth is receding from her at a speed nine-tenths the velocity of light, so she concludes that her earthbound brother is aging less rapidly than she is (Fig. 3). The question is: "If the twin in the spaceship were to travel for a few years, then turn around and return to earth, which twin would really be older?" The answer is that the twin in the spaceship ages less. Although at first glance the travel situation may appear to be symmetrical from both twins' points of view, from the earth or from the spaceship, there is, in fact, a key difference. The spaceship must eventually slow down, stop, turn around, and accelerate in the opposite direction to return to earth. In doing this the spaceship and its occupant experience large forces that the twin on earth does not feel. These forces distinguish one twin from the other. In contrast to the spaceship, the earth continues in its gentle way through space[1] without changing its velocity much (in comparison to the enormous velocity of light), so the twin on earth experiences only gentle forces. The spaceship's enormous *changes* in velocity are what distinguish it from the earth since it must slow down from nine-tenths the speed of light, reverse direction and speed up to nine-tenths the speed of light[2] before the twins can get together again to compare their ages. (Einstein's relativity theory shows that unless the twins are together in space, it is no easy matter to compare their ages with any meaningfulness.) The resolution of the twin paradox lies in the velocity changes, and physicists are now fairly confident that spaceship crew members could return younger than their

---

[1]The earth's speed in orbit around the sun is about 30 km/sec, far less than the speed of light (300,000 km/sec). In terms of this enormous velocity, the earth could be said to be standing still.

[2]The twin on the rocket ship could, if she liked, insist that the rocket moved with a constant velocity but that the rest of the universe suddenly accelerated so much that it caught up with her. However, she would then have to explain the forces she felt at the time the rocket ship reversed its course. For example, if she explained these forces as a sudden gravitational field that the rocket encountered, Einstein's theory can account for the effects of such a field; it does, in fact, give the result we have described above.

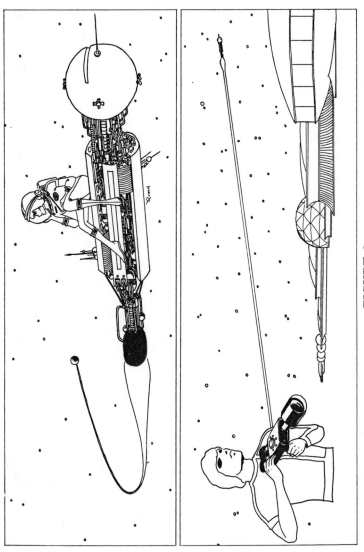

FIGURE 3
Time dilation: twin paradox.

A simple formula gives the amount of time dilation that occurs for a particle or for a system of particles traveling at a constant velocity v with respect to a fixed system:

$$\text{Time in fixed system} = \frac{\text{Time in moving system}}{\sqrt{1 - (\text{velocity})^2/(\text{speed of light})^2}}$$

In terms of symbols, this formula becomes

$$\text{Time in fixed system} = \frac{\text{Time in moving system}}{\sqrt{1 - (v^2/c^2)}}$$

This formula works only for a system moving with *constant* velocity, and it does not apply when a spaceship or a particle is accelerating, even though time dilation is occurring then, but of a different amount.

Suppose that a spaceship travels with a velocity equal to six-tenths the speed of light. How much time dilation occurs?

We note that in this example $v = \left(\dfrac{6}{10}\right)c$, or

$$\frac{v}{c} = \frac{6}{10} = 0.6$$

We can substitute this in our square root, so we have

$$\sqrt{1 - \frac{v^2}{c^2}} = \sqrt{1 - (0.6)^2} = \sqrt{1 - 0.36} = \sqrt{0.64} = 0.8$$

Then the formula which relates time in motion to time at rest tells us that

$$\begin{array}{l}\text{Time in fixed} \\ \text{system at rest}\end{array} = \frac{\text{Time in motion}}{\sqrt{1 - (v^2/c^2)}} = \frac{\text{Time in spaceship}}{0.8}$$

Thus 4 years spent in a spaceship traveling at six-tenths the speed of light is equivalent to the longer time on earth of

$$\text{Time at rest on earth} = \frac{4 \text{ years}}{0.8} = 5 \text{ years}$$

We have assumed the earth's velocity to be zero (the earth's speed in its orbit around the sun is much, much less than the speed of light).

The factor $\sqrt{1 - (v^2/c^2)}$ also appears prominently in the formula for the *kinetic energy* (KE) of a particle with mass M. For such a particle, the kinetic energy is[1]:

$$KE = \frac{M \times c^2}{\sqrt{1 - (v^2/c^2)}} - (M \times c^2)$$

For a proton with a speed equal to six-tenths the speed of light, the kinetic energy is

$$KE = \frac{M(\text{proton}) \times c^2}{\sqrt{1 - (0.6)^2}} - M(\text{proton}) \times c^2$$

$$= \frac{M(\text{proton}) \times c^2}{0.8} - M(\text{proton}) \times c^2$$

$$= 1.25 M(\text{proton}) \times c^2 - M(\text{proton}) \times c^2$$

$$= 0.25 M(\text{proton}) \times c^2$$

We can now answer the question of why a particle, or a system of particles, *with mass* can not travel at a speed equal to or greater than the speed of light. Consider the square-root factor $1 - (v^2/c^2)$. If the velocity of a proton

---

[1]This formula does *not* give the kinetic energy of a particle with *no* mass, like a photon or a neutrino; there is another formula for these particles. It can be shown by the mathematically inclined that the formula given above for the kinetic energy has the less complicated approximation

$$KE = \tfrac{1}{2} M \times v^2 \qquad \text{approximately}$$

in those cases where v^2 is very much less than c^2.

c, then the ratio v/c would become

$$\frac{v}{c} = \frac{c}{c} = 1$$

Inside the square root, we would have zero:

$$\sqrt{1 - \frac{v^2}{c^2}} = \sqrt{1 - (1)^2} = \sqrt{1 - 1} = \sqrt{0} = 0$$

But then the kinetic energy would be infinite since we have

$$KE = \frac{M \times c^2}{0} - (M \times c^2)$$

The formula gives us infinity for KE because zero divided into anything that is not zero gives infinity. Because there is never an infinite amount of energy to give to a proton or any other particle, we can not get particles with mass to travel at speeds equal to or greater than the speed of light. We can accelerate a particle with mass nearer and nearer the speed of light, but the particle will never quite reach the velocity c. As the particle's velocity v grows closer and closer to c, the particle's kinetic energy grows larger and larger. But this kinetic energy must come from converting some other kind of energy into the particle's energy of motion. No matter how much energy we use to accelerate the particle, it is not an infinite amount, so the particle will never reach the speed of light.

SUMMARY

Observations of systems of particles in motion shows that *time* in moving systems slows in comparison with time in a system of particles at rest. An example of this is the neutron. We observe a neutron to decay more slowly when the neutron is in motion than when it is at rest. This dilation or slowing of time is an essential part of relativity theory and has been verified experimentally. The time dilation effect increases steadily as the system's velocity approaches the velocity of light. This could someday allow fast-moving spaceship crews to return to earth younger than their twin brothers or sisters who stayed behind. The energy of motion of a particle with mass increases steadily as the velocity of the particle increases and would be infinite if the particle could reach the velocity of light. This is why particles with mass can not ever go as fast as light, though they can get nearer and nearer this speed if we use more and more energy to accelerate them.

QUESTIONS

1 The decay of a stationary pi-zero meson into two photons occurs in about 10^{-16} sec. If this pi-zero meson were moving at half the speed of light relative to us, would it appear to decay more quickly or more slowly?

2 If a space pilot made a journey at speeds near the velocity of light while her twin sister remained behind on earth, who would be younger when the pilot returned to earth? Why couldn't the pilot say that the earth had been moving near the speed of light relative to her, and so her earth-bound sister should be younger than she?

3 If Cyril and the Top Dog travel at eight-tenths the speed of light, how many years of their time are equivalent to 1 year of terminal time? If Cyril and the Top Dog travel at a speed 99 percent the speed of light, 1 year of their time will be equivalent to how many years of terminal time? What if Cyril and the crew are traveling with a speed 99.99 percent of the speed of light?

Helpful numbers:

$$\sqrt{1 - (0.99)^2} = 0.14 \qquad \sqrt{1 - (0.9999)^2} = 0.014$$

(*Ans.*: 0.6 years; 0.14 years; 0.014 years)

4 How much greater than its rest energy is the kinetic energy of a proton moving at 99.99 percent the speed of light? At 99 percent the speed of light? At eight-tenths the speed of light? (Use the square-root factors from Question 3.) (*Ans.*: 7.04 times; 6.14 times; 0.667 times)

7
photons

"Ah, the hero returns."

Zaki was never sure whether or not Hot Spur, Zenith Borg's chief adjutant, was putting him up. To be secure, Zaki decided to react straightforwardly.

"Yes, Commander, I'm back—another journey done. I see you're getting a trifle pudgy there." Cyril pointed a muscular thumb at Spur's midsection. Spur must be five hundred years old by now, Zaki thought. No wonder he needs a facelift. Another hundred and he'll have his pension.

"Well, we don't have the secret of eternal youth that you do, Captain. It's no easy job here at the Center, though I know you imagine us as having nothing better to do than shuffle tapes."

And build up your fat zor, Zaki said to himself.

"Well, Spur, I know I'm not considered competent to understand a lot of the higher thoughts around here. But since, as you perhaps are aware, I'm here to make a 4-B report, would you mind telling the I.E. I've arrived?"

"Not at all, Captain. In fact, I think you can go right in."

Spur pressed a significant button. Amid a soft tinkle of chimes, two carved panels slid back to reveal the power center of the Trust, where Zenith Borg raised her all knowing eyes to cast a questioning look at the captain as he entered.

Zaki strode into the room with a crease in his forehead. His huge hands trembled slightly as he strove to produce a confident greeting to Borg, the one person whom he truly feared and loved. He took a breath.

"All tubes still open, your Assurance."

"Cyril," she replied, "I'm afraid there's little time for your pleasantries. Lower my risk and tell me what happened on Sidney."

Though Cyril had never seen Zenith so tense, he found time to admire the way her mouth moved to produce the words that governed him.

"We went out there and were almost sucked into a black hole—they really do exist! We had to land on Sidney for repairs, and right away funny things began to . . ."

"In the name of the Great Insuror," Borg interjected, "Keep your report specific. What happened?"

"Well, to tell you the truth, the crew and I went out drinking. I know we shouldn't have done it, but it's legal there! You wouldn't believe all the things that are allowed on Sidney—they gamble, and they dissipate, and they dope up on . . ."

"Cyril, either you stop this borborygmus or I'll have you cleaning the tubes in the Dog's thrusters! You overdrank—then what?"

"Well, I passed out. I'm not sure where. When I woke up I was in the bowels of the Sidney Sumptuary Storehouse; that's a kind of museum they've got there. A strange man named Dr. Zed had me tied up and was telling me a lot of fancy hoo-hah about you and the Trust. He said we're doing things all wrong—that we should stop insuring and take more risks! But he wasn't a zoroo; some of what he said almost made sense. Not about insurance, I mean, but he seemed to know an awful lot. Like the magic sphere in the Dog's control room. Do you know what that's for?"

Zaki, who had been darting an occasional glance at Zenith as she sat listening impatiently to his report, now obtained her full attention. It was delightful, but a bit unnerving.

"So you found Zed!" she said. Her black eyes beamed with pleasure. "What did he ask you about me?"

Cyril's breath exploded in a mighty burst.

"You knew about Zed? And you didn't tell me when you sent me out to that danger area? Just what kind of a policy are you writing around here?"

It was Zenith Borg's turn to flash fire at her interlocutor.

"You fat-chested space monkey," she retorted, "Don't forget who you're talking to. I'm not Lisa Dalby or Julie Card. You're the Dog's captain because I judged you a good risk, and I'll replace you just as fast as that turns out to be a good policy. Now for the last time, stop your maundering and get it on. What did Zed say?"

Zaki's titanic anger began to cool even as Borg finished her question.

"He seemed to know an awful lot about you. I bet your own agent doesn't know more. He kept talking about how you studied old manuscripts when you first joined the Committee . . ."

"Yes? And did he know why I stopped?"

"He said the Committee decided it was a bad risk. Listen, Zenith, was there any truth in what he told me? Were our grandmothers and their grandmothers smarter than we are? How can there be any sense in going uninsured?"

Borg's brow furrowed slightly as she wondered how much Zaki could understand.

"Look at it this way, Cyril. Every generation preserves what it needs from its ancestors. We don't want what's dangerous or harmful. By now we've got ourselves well protected, and we don't have to waste time studying how to blow ourselves up or make new and ugly machines like they did in the old days. Of course that means we're

ignorant about some of the things our ancestors had or
knew, but we've got all the relics that we need, and
we keep the premiums coming in to guard against
misfortunes. And as for being uninsured: you tried
some of that on Sidney? Did you like it?"

"I was insecure," Zaki answered. "It was risky. Is that
how Zed wants everyone to feel?"

"Yes, I think so, but I'm not really sure. It's so long
since I knew him . . . ."

"So Zed was on Pilar, after all?" Zaki asked.

"Of course. He worked here at the terminal, on the
J ring, clearing canceled policies. It's tricky work because
you have to sort the useful information from all the
mistakes. We can't afford to have the Trust go bankrupt,
can we? But if people aren't insured, they can't function."

Cyril was impressed. This was the sort of stuff he
rarely thought about.

"Zed told me that he used to 'sift right down to the
Permids'."

"Permids! Did he say that?" Zenith burst out.

"Yes, he did, that's what is sounded like. When I
asked him what he meant he just laughed and said that's
where his little secret is. What was he talking about,
anyhow?"

Zenith Borg appeared lost in recollection. With her
eyes closed, the constant control always apparent in her
face seemed to relent for a moment. She refocused her
gaze on Zaki for a moment of real affection.

"Cyril, I'll tell you something that only I know. The
Permids are heaps of rocks on a planetary cinder called
Thurd. They were found by Trilling himself during the
great Era of Exploration. He thought they might be the
oldest known relics of civilization."

The double doors of Zenith Borg's inner office sprang
open, and in rushed Hot Spur, plainly in a snit.

"We're picking up an unauthorized transmission from
this room, Z. B. Is this man completely covered?"

Zaki resented the question. He had a sensitive resenter.

"What do you mean, covered, you uninsurable fancy-pants?" he demanded. "I've had every coverage a man can get."

"Hold on a minute, Cyril," Zenith interjected. "There may be something here that's very tricky. Spur, get a portable orgonometer in here on the double."

In a few minutes Spur was back, cradling a treasured relic that could detect radio emission at frequencies from ten to ten thousand megahertz. He twiddled several dials and announced his conclusion.

"Just as I thought, your Assurance. He's wearing a sender."

A look of disdain, coupled with a slight show of pity, spread over Zenith Borg's high cheeks.

"Cyril, take off that tunic. Right now."

"Oh, come on now, Zenith, what in cancellation do you think I'd . . . "

"Now, Zaki. Delete the fine print."

Cyril demurred no further. He raised his mighty arms and let the tunic fall to the floor. His gluteus maximus stood out like a knot of muscle. Spur brought the orgonometer closer.

"It's right here, Z. B. Must be hidden in his navel."

Zenith bent closer for a look at the rings of muscle that outlined Cyril's impressive gut.

"I don't see anything there—but what's this scar on his stomach?"

She pointed with an expressive finger at a barely visible line of stitch marks just above the premium number on Cyril's abdomen.

"That's it!" she cried. "Zed's quite a shrewd operator. Spur, get Doctor Cudaback in here, and tell him to bring his acupuncturist, that Ching Quat fellow. And hop to it, Spur, there's not a minute to lose."

Cyril Zaki sought to protest. "Come on, Zenith, don't be so ridiculous. Do you think Zed could have put a sender in my grinder without my knowing about it?"

"You can stake your massive policy on it, Cyril my

boy. But we'll say no more until we get a closer look."

As Borg spoke, she pressed numerous buttons on the teleconsole.

"Call the Committee members right away," she ordered. "Have them stand by for further communications."

Spur sprang through the doors again, announcing that the doctor and his needler had arrived from the G ring.

"Get them in here, man, and you, Cyril, lie down on that couch."

"Couldn't we check a little further, Zenith, I mean it's a little . . ."

Borg lost no time in answering Zaki's digressions.

"Proceed, if you please, Dr. Quat."

"Please call me Dr. Ching. Quat is the first name."

"What?"

"No, Quat."

"Now see here, my well-insured friend, I want you to get your needles into this man's arm without further hoo-hah, or I'll have you anesthetizing rocks on the back side of Hagedorn. Is that clear?"

"Yes, your Assurance."

Dr. Ching took Zaki's left arm in a sure grip and inserted a flexible steel needle into the inside of the elbow. Cyril managed to stay calm, averting his eyes as the doctor moved the implement with infinite tactility.

"We are ready now."

"All right, Dr. Cudaback, let's open him from his guggle to his zatch and see what's in there."

Borg indicated the suspect region in a clearing of wiry hair.

"Good risk," said the doctor. He produced a scalpel, requested that Cyril hold still, and made a ten-centimeter incision in Zaki's solar plexus. *Thank G. this acupuncture works,* Cyril thought. *I'm not really that tough.*

"Aha!" cried the doctor, and drew a shiny grey object from Cyril's midsection.

"Cancellation," whistled Spur.

"All right, doctor, stitch him up, and thanks. Cyril, we're leaving as soon as you can get your ship ready. Spur, tell the Committee I've got to handle this myself. It's me against Zed. No one else can do it."

"But Z. B., I mean your Assurance, you can't leave the terminal and go off into time dilation. I mean — it's just not done! You'll be gone for years and years, and who'll take care of the Trust?"

"The Committee will, Spur, Let them do something for a change. I tell you, this is too important for anyone else. Cyril, how are you feeling? I need you on this trip; you're the best pilot we've got. Don't be embarrassed just because Zed used you in his plans."

Cyril had begun to realize that he had been manipulated.

"You mean Zed sewed a sender inside me and then let me escape? Oh, junk!"

"Yes, I'm sure that's just what happened. He was waiting by the black hole for the Trust to send an investigator. Once you arrived, he captured you, wired you up, and then followed you back to Pilar, where he's been listening to your conversations to learn the secrets of the Trust. Now that he knows where the Permids are, he'll be on his way to Thurd. We've got to get there and stop him, so let's haul zors."

Ancient Egyptians considered light rays from the sun to be messengers from the sun god Ra. Ra in his infinite goodness warmed and brightened the days along the Nile. (Actually, the slaves who sweated and strained to build the great pyramid of Khufu probably wished that Ra could be covered in cow dung.) Four thousand years later, light still reaches us from the sun, though we do not see it in quite the same way as the ancient Egyptians.

Light is one form of electromagnetic radiation. Other common types of electromagnetic radiation are radio and television waves, radar, microwaves, infrared radiation, visible and ultraviolet light, and x-rays. Ordinary light, as well as all other forms of electromagnetic radiation, consists of elementary particles called photons. A photon has no electric charge and no mass. Photons do have energy of motion, but they have no energy of mass. A photon is identical to an antiphoton, and photons form a class of particles by themselves.

Photons have two main functions. First, they form light and all the other kinds of electromagnetic radiation. Secondly, they transmit the electromagnetic forces between particles with electric charge. We shall consider the photons' first function, that of making electromagnetic radiation.

All real particles that have no mass travel at the speed of light, 186,000 miles/sec, in empty space. We use the symbol c to designate this velocity. Such particles include neutrinos, antineutrinos, and photons. Since all electromagnetic radiation is composed of photons, it all propagates with the speed of light if there is no matter to obstruct its journey.

No real particle or antiparticle, and hence no matter or antimatter, can travel faster than the speed of light. In fact, particles that do have some mass must always travel at less than the speed of light and usually travel much more slowly than light.

The sun is about 93 million miles from the earth. A group of photons leaving the sun's surface as a beam of

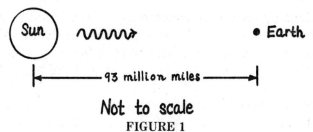

Not to scale
FIGURE 1
Photons leaving the sun take 500 sec to travel
the 93 million miles between the sun and the earth.

light bound for the earth, traveling at the speed of light, takes a time equal to 500 sec [or 8 minutes (min) and 20 sec] to reach the earth (Fig. 1). We can calculate this time by dividing the sun's distance from the earth by the speed of light:

$$\text{Time of travel} = \frac{93{,}000{,}000 \text{ miles}}{186{,}000 \text{ miles/sec}} = 500 \text{ sec}$$

If the sun suddenly were to stop producing light, we would still see sunlight for 8 min and 20 sec. This is the light that was already on its way to us before the sun stopped shining.

The star nearest us other than the sun is Alpha Centauri, about 24 trillion miles from the earth. Photons (light) that leave the surface of Alpha Centauri take almost 4 years to reach us. Stars' distances are often described in units of light years, which are the number of years it takes for the stars' light to reach us. If Alpha Centauri were to wink out today, it would be 4 years before we realized it. Some faraway stars are so distant that they no longer exist, although we are still receiving the photons they radiated long ago.

How can we imagine what a photon really is? One way to visualize a photon is to think of a tadpole or a snake moving from left to right across the page with the speed of light. As the tadpole moves, it constantly wiggles its body back and forth in a wavy motion, making a series

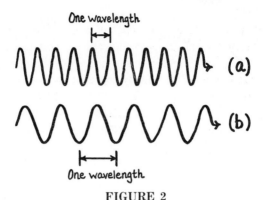

FIGURE 2
We can think of a photon as a series of undulations
or vibrations that travels at speed c.

of vibrations. The *frequency* of a photon is the number
of these vibrations that occur each second. In Fig. 2,
photon (a) has twice as many vibrations per second as
photon (b). This means that photon (a) has a frequency
twice as great as photon (b)'s.

A photon's energy of motion is proportional to its frequency of vibration. This is in accord with our intuition,
if we recall the analogy between a photon and a tadpole.
The faster a tadpole wiggles, the more energetic it
appears. The same is true for photons: Photons with more
rapid vibrations have greater frequencies and greater
energies than photons with less rapid vibrations. The
energy of motion of a photon equals the photon's frequency f times a constant of proportionality called h.
The constant h is the same everywhere and at all times.[1]
This means that in Fig. 2, photon (a) has twice as much
energy of motion as photon (b) because (a)'s frequency is
twice (b)'s. All this can be summarized as:

$$\text{Photon's energy of motion} = h \times f$$

---

[1]The constant h has the value 6.625×10^{-27} erg-sec. When h is multiplied
times a frequency, the units of frequency (per second) are changed into the
units of energy (ergs).

The distance from the crest of one wave of a photon to the crest of the next wave is called the photon's wavelength. This is also the distance over which a photon's wave pattern repeats itself. In Fig. 2, photon (a), with the larger frequency, has a wavelength one-half as great as the wavelength of photon (b), the lower (smaller) frequency photon. *Larger* frequencies (faster vibrations) mean *shorter* wavelengths because more oscillations can fit into a given length of tail (Fig. 2).

The wavelength of a photon is also identical to the distance a photon travels during the time it oscillates or vibrates back and forth once. To see this, we can move the photons in Fig. 2 past a fixed point, such as a firmly placed finger. To pass through one complete cycle of oscillation, the photon must move a distance equal to the distance between wave crests, which is one wavelength of the photon.

We can now derive a precise relationship between a photon's wavelength and its frequency by calculating the distance a photon travels during one complete oscillation or cycle of vibration in terms of the photon's *frequency*.

How can we perform this calculation? First, we relate a photon's frequency to the time that it takes for one oscillation to occur. Then we calculate how far the photon travels during this time interval. The photon's frequency, f, tells us how many oscillations (cycles) occur every second. One oscillation or cycle takes a time equal to 1 over the frequency. For example, a photon with a frequency of 1 million cycles per second (1,000,000 cycles/sec) takes one-millionth of a second (1/1,000,000 sec) for each of its oscillations. The distance that a photon travels during one oscillation or cycle is the speed of light c times 1 over the frequency. (We are using the general rule that distance equals speed times time.) The distance that a photon travels during one complete cycle is also equal to its wavelength λ, so we can equate the two quantities to obtain

$$\lambda = \text{one wavelength} = c \times \frac{1}{\text{frequency}} = \frac{c}{f}$$

Each type of electromagnetic radiation is distinguished by its frequency, or, alternatively, by the wavelength of its constituent photons. For example, the frequency of the photons that make up ordinary FM radio waves is 10^8 (100 million) cycles/sec, and these photons have a corresponding wavelength of 300 cm. On the other hand, the frequency of the photons that make up blue light is 10^{15} (1 million billion) cycles/sec, and their corresponding wavelength is 3×10^{-5} (thirty-millionths) of a centimeter. Figure 3 lists some common forms of electromagnetic radiation, together with the frequency and wavelength range of the photons that form the different types of electromagnetic radiation. Since the energy of each photon is proportional to its frequency (energy $= h \times f$), we could equally well say that each form of electromagnetic radiation is distinguished from the other types by the energy of its constituent photons.

Photons interact most strongly with objects that are about the same size as their wavelength. The wavelength of electromagnetic radiation is thus important because it tells us roughly what size of objects will interact strongly with the photons that comprise that particular form of electromagnetic radiation. For example, television waves have an average wavelength of 100 cm or about 3 feet (ft) (Fig. 3). Because the photons that make up television waves interact most strongly with objects 3 ft across, most television antennas are about this size.

Ordinary visible light has an average wavelength (5×10^{-5} cm) that is the size of several thousand atoms. The photons that comprise visible light interact strongly with the thousands of atoms in our eyes' retinas (these form the "rods" and "cones" in our eyes). We can now understand why we can not "see" elementary particles or their constituents. Ordinary visible light has an average wavelength 500 million times greater than a pro-

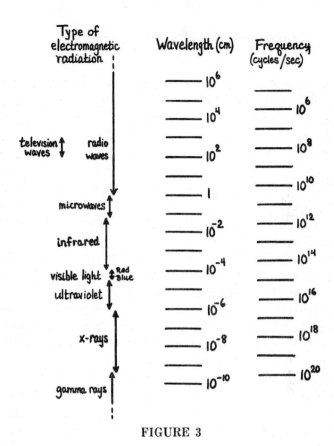

FIGURE 3

ton's radius (10^{-13} cm). The wavelength of visible light is so mismatched in size with protons that its photons do not interact with protons. Therefore, we can not see atoms or their constituents with ordinary light.

Infrared radiation consists of photons with wavelengths somewhat longer than those of the photons in visible light. The photons that form infrared radiation have wavelengths which are the size of a few thousand large molecules. When infrared photons reach our bodies, they have the right wavelength to jiggle groups of molecules gently and warm us.

Medical and dental x-rays have wavelengths about the size of atoms, and they interact strongly with atoms. In particular, they interact most strongly with the atoms of calcium in our bones when they are aimed at our bodies. The photons that form x-rays pass through our soft tissue and are absorbed when they interact with our bony structure. This leaves an outline of our bones on the x-ray film because the film is darkened by its exposure to the photons which pass through the fleshy tissue, but the film is not darkened where our bones have absorbed the x-ray photons.

What happens to those parts of our bodies that absorb the x-ray photons? We know from statistical studies of groups of people who have absorbed large amounts of x-rays that massive doses will produce cancer. We do not know the precise set of steps that leads from the interaction of one x-ray photon with one atom to one cancerous cell, and then eventually to many cancerous cells and a cancerous condition. However, some scientists believe that each x-ray photon has a miniscule possibility of initiating such a sequence of steps. To minimize the chances of such an occurrence, we should avoid absorbing lots of x-ray photons.

It is ironic that in some cases cancerous cells can be destroyed by x- or gamma-ray treatments. [Gamma rays are composed of high-energy photons with very short wavelengths (high energy means high frequencies or short wavelengths).] In these treatments, x- or gamma rays are focused on a particular growth (not on the entire body or all its bones) for a short time in an attempt to kill cancerous tissue.

The different *types* of electromagnetic radiation are distinguished from one another by the *energy* or *frequency* of their photons. The strength or *intensity* of a given type of electromagnetic radiation is proportional to the *number of photons* that form it. For example, an ordinary 100-watt light bulb emits 2×10^{18} photons every second. A more intense 200-watt bulb will emit twice as

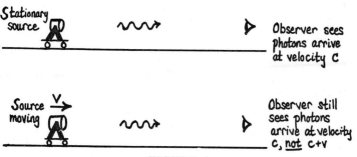

many, or 4×10^{18}, photons every second. The intensity of
light, and of other types of electromagnetic radiation, is a
measure of the number of photons in the radiation. When
visible light photons enter our eye, they interact with the
large groups of atoms and molecules that form the rods
and cones in the eye's retina. This disturbance on the
retina is then transmitted to our brain, where it is inter-
preted for us. This *physiological* process of vision relies on
two *physical* clues: the photons' wavelength (or alternatively, their frequency) and the number of photons in the
radiation. The different wavelengths of the photons are
translated into different colors by our brain, and the number of photons is translated into brightness, or intensity.

THE DOPPLER EFFECT

Photons have a strange property: they are always found
to be traveling at the speed of light (c), even when we
produce photons from a source that moves with respect
to the observer! If we mount a searchlight on a train
moving with speed v and watch the train come toward
us, we might expect to find photons arriving at speed
$c+v$ (Fig. 4). But they always arrive at speed c! Similarly,
photons observed to come from a searchlight receding
from us at speed v do not have velocity $c-v$ but just plain

FIGURE 4
Photons are always found traveling at velocity c
(186,000 miles/sec) through empty space *regardless*
of how fast the source of photons is moving.

156 c. This phenomenon, repeatedly verified, baffled scientists for years until Einstein made it part of his special relativity theory, which was based upon the observational fact that the speed of light is the same for all observers. The constancy of photons' speed holds true no matter how fast the observer is traveling with respect to the source of photons, and this fact must be part of any consistent explanation of how the physical world behaves.

However, the photons emitted from a photon source that moves with respect to an observer *do* appear to change as a result of their motion. We find no change in the photons' *speed* of travel, but we do measure a change in the photons' *energy.* When photons are emitted from a source that is moving *toward* us, they appear *more energetic* than they would if the source were stationary (Fig. 5). Conversely, when photons are emitted from a source that is moving *away* from us, the photons appear *less energetic* than they would if the source were stationary (Fig. 5). This phenomenon is known as the *Doppler effect.*

We can illustrate the Doppler effect by considering a hypothetical star that is far from the earth: Zed's star. When it is stationary with respect to us it radiates photons with a given energy (frequency) characteristic of

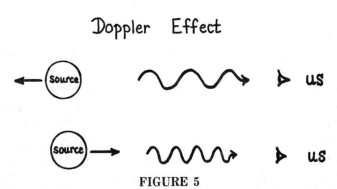

FIGURE 5

The relative motion between a source of photons and the observer changes the energy of the photons that we observe in comparison to the energy with which they were emitted by the source.

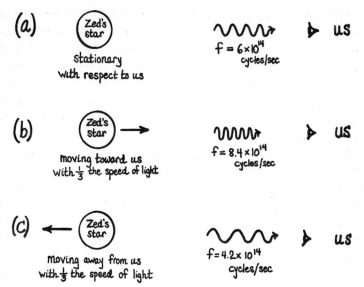

FIGURE 6

(a) A hypothetical star, called Zed's star, emits photons with a frequency of 6×10^{14} cycles/sec. If Zed's star is stationary with respect to us, then we observe photons with a frequency of 6×10^{14} cycles/sec. (b) If Zed's star moves toward us with one-third the speed of light, we see the photons as *more energetic*. The photons' frequency is greater and their wavelength is smaller. (c) If Zed's star is moving away from us with one-third the speed of light, we see the photons as *less energetic*, so the photons have a smaller frequency and larger wavelength than when they were emitted.

yellow light (frequency $= 6 \times 10^{14}$ cyles/sec), and we observe photons with this frequency (Fig. 6a). If for some reason Zed's star started to move toward us, or if we moved toward it, we would now detect its photons as having a *greater energy* (frequency) than we did when the star was stationary with respect to us (Fig. 6b). The faster the star moves toward us, the greater will be the increase in the energy (frequency) of the photons we observe. If Zed's star were approaching us with a velocity one-third the speed of light, the increased frequency of the photons that we detect would be 8.4×10^{14} cycles/sec. We would then see photons from Zed's star as blue instead of yellow light. On the other hand, if the star were moving away from us, or if we were moving away from it, we would find that we observe its photons to have *less energy* and thus a smaller frequency than when the star was stationary with respect to us. The faster the star recedes from us, the less energetic (lower frequency) its photons will appear to us. If Zed's star were moving away from us with one-third the speed of light, the decreased frequency of its photons would be 4.2×10^{14} cycles/sec. This would cause us to observe the photons as red light, even though they were emitted with the frequency of yellow light (Fig. 6c).

The important thing to remember about the Doppler effect is that relative motion of the source *away* from us makes the photons *redder* (lower frequencies), and motion *toward* us makes the photons *bluer* (higher frequencies). The Doppler effect depends upon the relative motion between the source of photons and the observer at the time when the photons were emitted, and it enables us to determine this relative motion *if* we know the actual frequencies of the photons when they were emitted. The change in the photons' energy gives the amount of the relative motion of the photon source toward or away from us.[1] Photons always travel at the same speed through space. When we observe a photon, we find that it always travels at the same speed, but its measured energy depends on its energy at emission and also on the motion of the photon source relative to the observer.

---

[1]The formulas for the change in the photon's energy is

$$\frac{\text{Observed energy}}{\text{Original energy}} = \sqrt{\frac{1 - (v/c)}{1 + (v/c)}}$$

where velocity v toward an observer is taken as negative and velocity away from an observer is taken as positive. For v much less than c, this formula gives approximately

$$\frac{\text{Observed energy}}{\text{Original energy}} = \text{(app.) } 1 - \frac{v}{c} \qquad \text{if } v \text{ is much less than } c$$

We can now describe the "quantized" energy levels of a hydrogen atom and discuss how a hydrogen atom interacts with photons. In Chap. 2 we presented a model for a hydrogen atom in which the electron orbits the proton in a circle (see Fig. 3, Chap. 2). Quantum-mechanical calculations[1] show that in a hydrogen atom, and in all other atoms, only certain, definite values are possible for the average distance of the electron from the proton. In terms of our picture of circular electron orbits, the only values allowed for the radius r of the orbits are those given by the formula

$$r = N^2 \times (0.53 \times 10^{-8} \text{ cm})$$

where N is one of the integers, so $N = 1, 2, 3$, and so on. *No* other orbits are possible for the electron in a hydrogen atom under the rules of quantum mechanics, which govern the behavior of tiny systems like atoms. The smallest orbit for the electron in a hydrogen atom is the one with $N = 1$, when $r = 0.53 \times 10^{-8}$ cm; the next smallest orbit has $N = 2$ and $r = 2.12 \times 10^{-8}$ cm; and so forth for larger N's.

For each particular (quantized) value of the electron orbit's radius r, the hydrogen atom has a specific energy, called a quantized energy level. To see this, we consider the total energy of a hydrogen atom. This total energy is the sum of two kinds of energy. First, there is the energy of mass of the proton and the electron, which depends only on the masses of the proton and the electron. Secondly, there is the *binding energy* of the atom, which measures how strongly the electron and proton are bound

---

[1]Quantum-mechanical considerations also show that we can not locate the electron with complete accuracy within an atom, so we must talk about the *average* distance of the electron from the proton throughout the orbit rather than about an exact distance of the electron from the proton at every instant.

by electromagnetic forces. This *binding energy* is a concept that applies only to the entire atom considered as a system, and it has no meaning in terms of the proton or the electron considered separately. However, the calculation of the binding energy does allow for whatever kinetic energy the electron has in its orbit, so the total energy of an *atom* is correctly given by the atom's energy of mass plus its binding energy.

The energy of mass of the proton and electron in a hydrogen atom is

Energy of mass = [mass(proton) × mass(electron)] × c^2

because the energy of mass is always the mass times the speed of light squared.

The binding energy BE depends only on the distance between the electron and the proton. The binding energy takes into account the fact that the electron orbiting the proton wants to escape, but the electromagnetic attraction of the proton for the electron keeps the electron in orbit. The formula for the binding energy of a hydrogen atom is

$$\text{BE} = \frac{K \times (e) \times (-e)}{2 \times r}$$

or, since $(1) \times (-1) = -1$,

$$\text{BE} = -\frac{K \times (e) \times (e)}{2 \times r}$$

where K is the constant of proportionality in Coulomb's law of electromagnetic forces (see p. 91), e is the electric charge of the proton, $-e$ is the electric charge of the electron, and r is the quantized radius of the electron's orbit.[1]

---

[1] Recall that the numerical values of K and e are

$$K = 9 \times 10^{18} \frac{\text{gm} \times \text{cm}^3}{\text{coul}^2 \times \text{sec}^2} \qquad e = 1.6 \times 10^{-19} \text{ coul}$$

The negative binding energy indicates that if we wanted to separate completely the proton and the electron from each other, we would have to kick them apart with an amount of kinetic energy equal to the binding energy.[1]

Even though the electron in orbit has some kinetic energy, it is not enough to overcome the electromagnetic forces that bind it in orbit around the nucleus, and therefore the atom's binding energy is a negative number. A binding energy of zero occurs when the radius of the electron's orbit gets infinitely large, so the atom is entirely unbound. Since the binding energy is inversely proportional to r, the orbital radius, we can see that an electron in the smallest orbit (smallest N value) has the most negative binding energy. Thus electrons are most tightly bound in atoms when they are in the smallest possible orbit.

If we substitute the numerical values for K and e and the quantized values for the radius r ($r = N^2 \times 0.53 \times 10^{-8}$ cm) into the formula for the binding energy, we find that

$$BE = -\frac{21.8 \times 10^{-12} \text{ erg}}{N^2}$$

where $N = 1, 2, 3$, and so on. The lowest (most negative) binding energy occurs when $N = 1$, and then $BE = -21.8$

---

[1] The binding energy of an atom is itself the sum of two energies. First, there is the electron's kinetic energy KE which is always positive. Secondly, there is the negative electromagnetic potential energy PE of attraction between the proton and the electron. The binding energy is negative when the (negative) attractive potential energy is a larger number than the positive kinetic energy, and the electron is then confined to one particular region of space, namely, its orbit around the proton. In general, the binding energy of a hydrogen atom is

$$BE \text{ (hydrogen atom)} = \underbrace{\frac{1}{2} \times M(\text{electron}) \times v^2}_{KE} + \underbrace{\frac{(e) \times (-e)}{r}}_{PE}$$

where v is the electron's orbital velocity. For the hydrogen atom with circular orbits the kinetic energy is equal to $[(e) \times (e)]/(2 \times r)$, so the binding energy is

$$BE \text{ (hydrogen atom)} = -\frac{(e) \times (e)}{2 \times r}$$

162 × 10⁻¹² ergs. This corresponds to $r = 0.53 \times 10^{-8}$ cm for the electron's orbit (Fig. 7). The binding energy for an atom with the electron in the $N = 2$ orbit is -5.45×10^{-12} ergs. This corresponds to $r = 2.12 \times 10^{-8}$ cm for the electron's orbit (Fig. 7). Given the chance, a hydrogen atom, and other atoms as well, *seeks to be in the state with the lowest binding energy* or an $N = 1$ state for its electron orbit.

The total quantized energy of a hydrogen atom in the Nth state is

$$\text{Total energy} = [M(\text{proton}) + M(\text{electron})] \times c^2 - \frac{21.8 \times 10^{-12} \text{ erg}}{N^2}$$

where $N = 1, 2, 3$, and so on.

We can now describe the interaction of photons with atoms. Only photons with certain, definite quantized

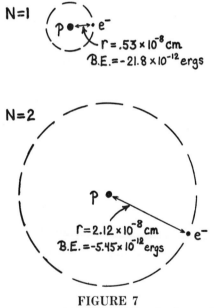

FIGURE 7

The $N = 1$ state of the hydrogen atom has the lowest binding energy (-21.8×10^{-12} ergs) and the smallest radius ($r = 0.53 \times 10^{-8}$ cm) for the electron's orbit around the proton. The $N = 2$ state has the next lowest binding energy and the next smallest orbit.

energies or frequencies will interact with a hydrogen atom. This behavior reflects the fact that the energy of a hydrogen atom can have only certain quantized values.

Let us consider the "excitation" of a hydrogen atom by a photon. Imagine a hydrogen atom in its $N = 1$ lowest energy state, and a photon about to collide with it (Fig. 8). During the collision, the electron absorbs the photon and becomes *more energetic*. After the collision, the hydrogen atom is in the excited $N = 2$ state (Fig. 9). To understand this process, we want to calculate the definite energy of the photon that is needed to excite a hydrogen atom from its $N = 1$ to its $N = 2$ state.

We use the principle of energy conservation: the total energy before equals the total energy after. The *total energy before* is the sum of the photon's energy ($h \times f$) and the energy of the hydrogen atom in its $N = 1$ state, or

$$\text{Total energy before} = (h \times f) + [M(\text{proton}) + M(\text{electron})]$$
$$\times c^2 - 21.8 \times 10^{-12} \text{ erg}$$

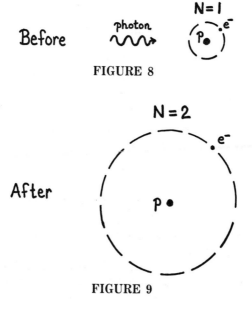

Before   *photon*   N = 1

FIGURE 8

After   N = 2

FIGURE 9

The *total energy after* is just the energy of the hydrogen atom in its excited $N = 2$ state, or

$$\text{Total energy after} = [M(\text{proton}) + M(\text{electron})] \times c^2$$
$$- 5.45 \times 10^{-12} \text{ erg}$$

We can equate the total energy before and the total energy after:

$$\text{Total energy before} = \text{Total energy after}$$

or

$$\{(h \times f) + [M(\text{proton}) + M(\text{electron})] \times c^2 - 21.8 \times 10^{-12} \text{ erg}\}$$
$$= \{[M(\text{proton}) + M(\text{electron})] \times c^2 - 5.45 \times 10^{-12} \text{ erg}\}$$

Since the energy of mass $[M(\text{proton}) + M(\text{electron})] \times c^2$ cancels on both sides of the equals sign, our equation becomes

$$h \times f = -5.45 \times 10^{-12} \text{ ergs} + 21.8 \times 10^{-12} \text{ ergs}$$

Thus the photon's energy $(h \times f)$ is

$$h \times f = 16.35 \times 10^{-12} \text{ erg}$$

This calculation tells us the frequency of a photon that can excite a hydrogen atom's electron from its $N = 1$ to its $N = 2$ orbit. To produce this transition, we need photons with this definite frequency. Photons with other frequencies will not excite a hydrogen atom's electron from its $N = 1$ to its $N = 2$ state.

We now consider the alternate situation to the one just described. An excited hydrogen atom emits a photon and goes to a less excited (less energetic) state. Consider a hydrogen atom in an excited $N = 3$ state (that is, its electron is in the third-smallest orbit).[1] Figure 10 shows

---

[1]The lowest-energy ($N = 1$) state of an atom is called the *ground state*. States with N greater than 1 are called *excited states*. The excited states all have more energy than the ground state of the atom.

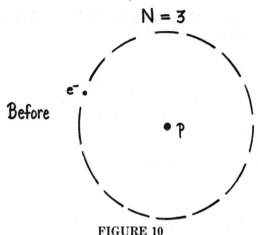

FIGURE 10
A model for a hydrogen atom in the $N = 3$ excited state.

a model of this situation. After a short time (less than one-millionth of a second), the electron emits a photon and falls back into the second ($N = 2$) orbit, so the atom is in the $N = 2$ excited state (Fig. 11). The electron might also fall back directly to the $N = 1$ state, but we consider the other possibility here.

We can determine the energy of the photon emitted when the electron falls from the $N = 3$ to the $N = 2$ state by using the principle of *energy conservation*. The total energy before is the energy of the hydrogen atom in its $N = 3$ excited state, or

$$\text{Total energy before} = [M(\text{proton}) + M(\text{electron})] \times c^2$$
$$- \frac{21.8 \times 10^{-12} \text{ erg}}{(3)^2}$$
$$= [M(\text{proton}) + M(\text{electron})] \times c^2$$
$$- 2.42 \times 10^{-12} \text{ erg}$$

The total energy after is the photon's energy ($h \times f$) plus the energy of the $N = 2$ state of the hydrogen atom, or

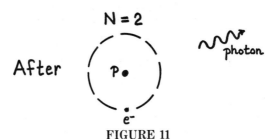

FIGURE 11

After a short time, the atom can emit a photon and
pass to the less excited (lower energy) $N = 2$ state.

Total energy after $= (h \times f) + [M(\text{proton}) + M(\text{electron})] \times c^2$
$$- 5.45 \times 10^{-12} \text{ erg}$$

We now equate the total energy before and the total
energy after, and we have

$$\{[M(\text{proton}) + M(\text{electron})] \times c^2 - 2.42 \times 10^{-12} \text{ erg}\}$$
$$= \{(h \times f) + [M(\text{proton}) + M(\text{electron})] \times c^2 - 5.45 \times 10^{-12} \text{ erg}\}$$

Since the energy of mass $[M(\text{proton}) + M(\text{electron})] \times c^2$
cancels on both sides of the equals sign, our equation
becomes

$$5.45 \times 10^{-12} \text{ erg} - 2.42 \times 10^{-12} \text{ erg} = h \times f$$

or the photon's energy $(h \times f)$ is

$$h \times f = 3.03 \times 10^{-12} \text{ erg}$$

All atoms interact with photons in the same way that
hydrogen atoms do. A photon with the right frequency
can excite any atom's electron into a higher energy state.
Conversely, an excited electron in an atom can emit a
photon and drop back into a less excited state.

The binding energy for the outermost electron of an
atom with atomic number Z is

$$BE = -\frac{Z^2 \times (21.8 \times 10^{-12} \text{ erg})}{N^2}$$

To excite the outer electron of a helium atom ($Z = 2$) from its $N = 2$ to the $N = 3$ state, a photon would need four times ($Z^2 = 2^2$) the energy of a photon that would produce the same change in a hydrogen atom.

We can now understand why x-ray photons can be chosen to interact with the calcium atoms ($Z = 20$) in our bones and not with the atoms (average $Z = 6$) that form our flesh. These fleshy parts consist mainly of molecules that are a mixture of hydrogen, carbon, and oxygen and have an average Z of 6 for the atoms in them. Let us suppose that for both the calcium and the fleshy atoms we wanted a photon to produce a change from the $N = 3$ to the $N = 4$ state of the atom. The calcium atoms would need a photon with $Z^2 = 20^2 = 400$ times the energy needed for a similar change in a hydrogen atom. The fleshy atoms need a photon with only 36 times more energy than that needed for a hydrogen atom. Thus if we produce a high-energy photon, we can arrange for the photon to excite the bony and not the fleshy atoms of our body.

We have used our model for the hydrogen atom to show the way that photons can interact with atoms in general. A photon with the right frequency can excite any atom's electrons into a higher energy state. Conversely, an excited electron in an atom can emit a photon and drop into a less excited state, if that state is not already filled with the maximum allowed number of electrons. Each photon emitted in this manner has a definite energy, frequency, and wavelength that reflect the kind of atom (because the energies depend on the atomic number Z) that emitted the photon and the N numbers of the energy levels involved in the production of the photon.

When we study celestial objects, such as the sun and other stars, that produce photons, we can analyze the number of photons produced with different frequencies to determine what kinds of atoms form the objects. Each kind of atom (hydrogen, helium, carbon, and so on) will produce photons with different characteristic frequencies. Some celestial objects, such as cold clouds of diffuse gas, do not produce any photons themselves; instead, they remove photons from a source of light that happens to lie behind them by means of the absorption process which we described for a hydrogen atom. Again, by studying the frequencies of light removed (absorbed) from the source's light rays, we can find out the kinds of atoms in the absorbing material.

If a photon with an energy greater than the binding energy of a hydrogen atom hits the atom, the photon can knock the atom completely apart, reducing the atom's binding energy to zero and, in addition, giving the proton and the electron some additional energy of motion. This process, called *photoionization*, can occur for any kind of atom when a photon hits the atom with enough energy to knock off one or more of its electrons. An atom with one or more of its electrons missing is said to be "ionized." Such an ionized atom will tend to attract electrons because it now has a net positive electric charge. The energy needed to ionize an atom is much less than the energy needed to fuse nuclei or to break nuclei apart. This is why atoms are being formed and broken apart in many places (in our own atmosphere, for example), but nuclear fusion and nuclear breaking occurs only where we have tremendous energies, as in the centers of stars or hydrogen bombs.

SUMMARY

Photons are the elementary particles that form light and all other forms of electromagnetic radiation, such as radio, television, infrared, visible light and ultraviolet waves and x- and gamma rays. A photon can be thought

of as a wiggling tadpole characterized by a wavelength,
frequency, and energy. The more energetic the photon, the faster it wiggles and the higher its frequency. Photons always travel at the same speed in empty space. For any photon the wavelength times the frequency is equal to this constant speed c (300,000 km/sec or 186,000 miles/sec). The photon's energy is directly proportional to its frequency (energy $= h \times f$, where h is another constant). A photon can have any energy, frequency, or wavelength, but it always obeys the relations given above. Observations of photons show that their speed is constant *regardless* of any relative motion between the source of photons and the observer. However, if a photon is produced by a source moving toward us, we see the photon to have *more* energy than it did when it was emitted, and if the source moves away from us, we see the photon with *less* energy than it had when emitted. This is called the Doppler effect.

In an atom, an excited electron can lose energy by emitting a photon. In this process, the electron drops from a larger to a smaller atomic orbit while the photon carries off the extra energy. Conversely, an atom's electron can be excited (gain energy) by absorbing a photon. In this case, the electron goes from a smaller to a larger orbit because the photon donates its energy of motion to the electron.

QUESTIONS

1 What kind of particles make gamma rays? x-rays? What kind of particles form rays of visible light? Radio waves?

2 What kind of particles form the nuclei of atoms?

3 How fast does a photon travel?

4 If a star is 1.86×10^{14} miles from us, how long does it take for the star's light to reach us? (*Ans.*: 10^9 sec, or 31 years and 7 months)

5 Suppose that a woman living on Mars sends a radio message to her son on earth. If Mars is 37 million miles from earth, how long does it take for her to hear her son's reply to her questions? (*Ans.*: 3 min 20 sec each way)

6 The cells in a person's body are about 10^{-3} cm in diameter. What frequency of electromagnetic radiation would interact strongly with these cells? (See Fig. 3 for the frequency of different type of electromagnetic radiation.) (*Ans.*: $f = 3 \times 10^{13}$ cycles/sec)

7 The energy output of a 100-watt light bulb is about 10^9 ergs/sec. Suppose that all this energy were emitted in the form of photons (light) with a frequency of 6×10^{14} cycles/sec. What is the energy of one of these photons? How many of these photons, approximately, are emitted every second by the 100-watt light bulb? (*Ans.*: 4×10^{-12} erg, 2.5×10^{20} per second)

8 Do the orbits of an electron in a hydrogen atom become larger or smaller if the atom goes to a higher energy state?

9 What is the frequency of a photon needed to excite a hydrogen atom in its $N = 1$ to its $N = 2$ state? (*Ans.*: 2.46×10^{15} cycles/sec)

10 Suppose that a hydrogen atom in its $N = 2$ state passes to its $N = 1$ state, emitting a photon as it does so. What is the wavelength of this photon? (*Ans.*: 1.22×10^{-5} cm)

11 If a very energetic photon hits a hydrogen atom, sometimes it will ionize the atom by completely separating the electron from the proton so that the electron is no longer in orbit. In this case, the atom's binding energy is zero, and the energy of the final proton-electron system is equal to their energies of rest mass plus the kinetic energies of the proton and electron. What energy does a photon need to ionize a hydrogen atom in its $N = 1$ state and produce a motionless proton plus an electron with 10^{-12} ergs of kinetic energy? (*Ans.*: 22.8×10^{-12} erg)

8
lumps
of matter

Borg and Zaki glided over the terminal and headed for the spaceport. Cyril's huge hands trembled slightly as he used the familiar controls to direct the four thrusters of the aircar. His usual feelings of mastery were temporarily overcome by his realization that his knowledge was indeed limited. Take this aircar, for instance, he reminated. Air forced from underneath pushes on the ground, right. Now the higher the pressure, the smaller the volume. Or was it the higher the pressure, the higher the temperature? Zaki recalled the space-training classes he'd muddled his way through. Those formulas, so easily forgotten. PeeVee equals Arty, he thought. Where did that come from? Twinkle twinkle little star, power equals eye squared are. Or was it little star up in the sky, power equals are squared eye? What about this aircar, then? In the dim recesses of his mind, Cyril could picture the nozzles underneath his vehicle forcing air downward. It seemed that the closer the aircar came to the ground, the more upward push he felt. How can it do that, Cyril wondered. He imagined a blob of air shot at the earth and bouncing off. Where did the pressure come from? Let's see, he said to himself, the atoms move around faster if they're pushed around faster . . .

"Great Trust, Zaki," Zenith Borg yelled at him, "Can't

you get there any faster? And watch those pylons a little more carefully, you cornerclipping spacebunny."

"Aagh," Cyril rebutted, but the force of his answer was lost as the spaceport came into view. Zaki whipped the aircar into a wingover turn and set his delicate cargo down softly by the Top Dog's service module. Waiting for him were the rest of the crew, hastily recalled from their arrival celebrations. Each knew the dangers of reentering time dilation so quickly after touchdown, but their faces presented a look of tense eagerness to their captain and their commander. Boyer ventured a light-hearted remark.

"Just up down up down, eh, captain? Where will it all end?"

Zenith Borg cast upon Boyer a cold eye of disdain.

"Can't we leave this zoroo behind?" she asked.

"Who'll read the E meter?"

"All right, then, let's blast off, or whatever you call it," ordered the I.E.

"We call it 'making the big E'."

"Well, make me a biggy then, but do it. You're making me insecure about your competence, I can tell you that."

17

Although it is interesting to know how matter behaves at the elementary-particle level, most of our experience in life deals with the behavior of large groups of particles. In this chapter we shall discuss matter that is composed of huge aggregates of particles. The particles can be elementary particles like protons and electrons, atoms like hydrogen or helium atoms, or molecules like water or carbon dioxide. In general, we shall be discussing matter that consists of at least 10^{20} particles, so we shall describe matter in bulk quantities.

Most matter can exist in three possible states: solid, liquid, and gaseous. For example, water can exist as a solid (ice), a liquid (ordinary water), or a gas (steam).

The particles that form a solid tend to vibrate or oscillate around relatively fixed positions (Fig. 1). In most solids, electromagnetic forces constrain the particles to their almost fixed positions. In some solids the atoms or molecules are arranged in regular arrays or lattices (Fig. 2). These solids are usually called crystals; quartz is an example of a crystalline solid.

The energy of a particle in a solid has two parts: the particle's energy of motion, which is the kinetic energy

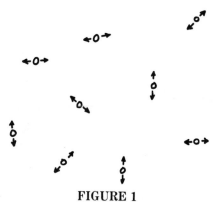

FIGURE 1

The particles that form a solid are constrained into almost fixed positions and vibrate only slightly around these positions.

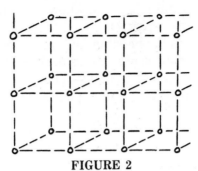

FIGURE 2
The particles in a crystalline solid are arranged
in a regular lattice with a repeating structure.

related to the particle's vibrations, and its binding energy.[1]
The binding energy reflects the fact that electromagnetic
forces hold the particle in an almost fixed position in the
solid.

In a liquid, large groups of particles clump together,
but these clumps are free to move with respect to one
another. Furthermore, particles in one clump can be
exchanged with particles in another clump (Fig. 3). The
energy of the particles in a liquid has two parts: an
energy of motion (kinetic energy) and a binding energy.
The binding energy reflects the fact that the particles are
bound into clumps of particles through electromagnetic
forces. As a general rule, the average energy of particles
in the liquid state is larger than the average energy of
these same particles in the solid state.

In a gas, the particles are free to move in all directions,
independent of one another. The particles—elementary
particles, atoms, or molecules—that make up a gas interact with each other only briefly, when they collide. Essentially all the energy of the particles in a gas is in the form

---

[1]Strictly speaking, we should also include the particle's energy of mass as
part of its energy. We do not include the energy of mass for convenience
because this energy of mass remains unchanged for large groups of particles
in the solid, liquid, or gaseous state, but the particles' average kinetic-plus-
binding energy does change for each of these states.

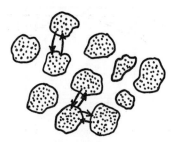

FIGURE 3
In a liquid, the particles form large clumps
that move rather freely with respect to one another.
Particles can be exchanged between the clumps.

of energy of motion (kinetic energy) because there is no binding energy. The average energy of particles in the gaseous state is larger than the average energy of these same particles in the liquid or solid state.

The temperature of matter composed of large aggregates of particles is a measure of the average energy of each particle. The higher the temperature of a piece of matter is, the larger the average energy of its constituent particles. When two pieces of matter are close to one another, they will tend to reach the same temperature. To do this, the pieces of matter transfer energy until their temperatures are equal. In this process, collisions among the constituent particles transfer energy from the hotter to the cooler piece of matter until the average energy of each of the particles that comprise the two pieces is the same.

For matter in the gaseous state, the relationship between the temperature of the gas and the average energy of its constituent particles is simple: The gas temperature is directly proportional to the average energy of motion of each of its constituents. (This relationship is simple because there is no binding energy of the particles.) If a gas is confined in a fixed volume by a container, the pressure that the gas exerts on the container will increase if the gas temperature rises (Fig. 4) because the particles

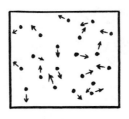

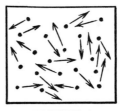

FIGURE 4
If a gas is confined inside a container, the pressure
of the gas will rise if the temperature increases
because the particles each have more energy of motion
and exert more force on the walls of the container.

move faster and exert more force on the container as they
collide with its walls. If a gas is not held inside a container but is free to expand, a rise in the gas temperature
will increase its volume. The particles that form the
gas will move farther apart as the temperature rises
(Fig. 5). The increased separation between particles
occurs because the particles move faster with increased
temperature and knock each other farther apart through
their collisions. The farther the particles (atoms or molecules) are separated in a gas, the less dense the gas is
because the density is defined as the amount of mass per
unit volume.

Stars are composed of gas that consists mainly of protons, each of which has a positive electric charge, and
electrons, each of which has a negative electric charge;
most of the mass is in the protons. A gas made of charged
particles is called a *plasma*. In a star like the sun, the
plasma (gas) of protons and electrons has an average
density of approximately 1 gm/cm^3, which is about the

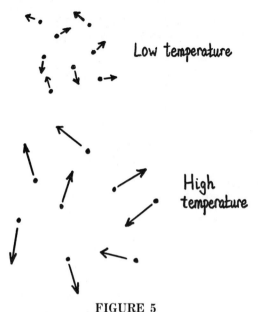

FIGURE 5
If the particles in a gas are not confined
in a container, then an increase in the gas temperature
will increase the separation between the particles.

density of water on earth. For comparison, we can consider the density of hydrogen gas here on earth. This gas
is also composed essentially of protons and electrons,
bound into hydrogen molecules. Each hydrogen molecule
consists of two atoms of hydrogen. Recall from Chap. 2
that the mass of a hydrogen atom is almost equal to the
mass of its central proton (see page 32) since the atom's
electron is 1,836 times less massive than the proton. At a
temperature of 0° Centigrade (0°C) and at normal atmo-
spheric pressures, the density of hydrogen gas is about
1/2,000th of a gram per cubic centimeter. This is 2,000
times less than the density of the protons (hydrogen gas)
at the center of a star like the sun. Gas in the sun can be
far denser than gases on the earth because the gravitational forces that hold the gas atoms together are much

stronger on the sun than on the earth. The strong solar force of gravity can squeeze the hydrogen gas to a relatively high density even though the material is hot enough (thousands or millions of degrees) to remain in the form of a gas, with no particles held in place as in a liquid or solid.[1]

All matter, if left undisturbed, will gradually share its energy with its surroundings until its temperature is equal to that of the surroundings. The matter can reach this equilibrium state by emitting and absorbing photons, each of which carries some energy, as described in Chap. 7. We shall describe this process for a gas made of helium atoms, but the process is similar for any kind of matter.

As the helium atoms that form the helium gas move around, they collide briefly with one another. Sometimes during one of these collisions an electron in one of the atoms will be excited and gain energy. This energy is gained by reducing the energy of motion of the original colliding helium atoms. A typical collision will take an electron in a helium atom from the $N = 1$ to the $N = 2$ or $N = 3$ orbit, thus raising the electron to a higher energy level. After the collision, the excited electron loses energy by dropping back to its original, unexcited $N = 1$ state as the atom emits a photon. The photon carries off the energy of excitation in the form of

---

[1] If the sun contracted without losing any mass to the point where it had a radius of 2 miles, it would become a black hole, as discussed in Chap. 1. As a result of such a contraction process, the sun's density would increase 8×10^{15} times because particles would be packed together much more tightly than they are inside the sun. Matter compressed to such an enormous density would behave in completely different ways from ordinary matter. The elementary particles—protons, neutrons, electrons, and so forth—which form the usual building blocks of matter on the earth and within stars would then be squeezed together at such enormous pressures that the ordinary particles would combine to form new, "exotic" particles. The density could increase without any known limit because the physics which we know today predicts that a piece of matter which has become a black hole will continue to shrink forever into some barely understood, superdense configuration.

energy of motion. This process occurs over and over among the colliding helium atoms. The net result of these collisions is that the gas radiates photons. These photons each carry off energy, and so the helium gas can slowly lose energy. This process continues until the gas loses enough energy so that its temperature is equal to that of its surroundings.

In the reverse process, a gas made of helium atoms can also absorb photons from its surroundings. If photons enter the gas with the right energies to be absorbed by the helium atoms (for example, by producing a transition from the atoms' $N = 1$ to the $N = 3$ state), they can give energy to the atoms that can appear as additional energy of motion.

If we allow sufficient time for the helium gas to reach equilibrium with its surroundings, the atoms in the gas will emit the same number of photons per second as they absorb, and thus there will be no net flow of energy between the gas and its surroundings. The tendency for matter to come into equilibrium (energy balance) with its surroundings is universal, and if enough time passes, all parts of the universe will eventually reach the same temperature with the same average energy per particle.

However, many parts of the universe (stars, for instance) are now much hotter than their surroundings. These regions are trying to reach equilibrium with the rest of the universe by emitting far more photons per second than they absorb. If this process went on long enough, the stars would reach a balance between the number of photons they emit and the number they absorb each second. This would occur at a very low temperature because most of the universe is empty.

★　★　★　★　★　★　★　★　★　★

We have described how a gas made of helium atoms will emit and absorb photons as it tries to reach an equilibrium with (have the same temperature as) its surroundings; this description would apply to any form of matter. A system of particles that has reached equilibrium with its surroundings is called an "ideal radiator." Such an ideal radiator will constantly exchange photons between itself and its surroundings to maintain this equilibrium, though there will be no *net* flow of energy in either direction. (Only at a temperature of absolute zero would no photons be emitted or absorbed.) An ideal radiator with a particular temperature emits (and absorbs) a specific number of photons per second, each with a specific energy. For example, consider an ideal radiator at a temperature of 3000° above absolute zero. Every second it might emit 10^{10} photons, each with an energy of 0.8×10^{12} ergs; $\frac{1}{2} \times 10^{10}$ photons, each with an energy of 0.4×10^{12} ergs; and $\frac{1}{4} \times 10^{10}$ photons, each with an energy of 0.2×10^{10} ergs. The photons that are emitted and absorbed by an ideal radiator at any given time have a characteristic distribution of energies, shown in Fig. 6. All ideal radiators will emit such a characteristic distribution of photon energies. If we plot the number of photons emitted at each different energy from an ideal radiator, we always find a curve with a peak value at some particular energy and a more rapid falloff toward low rather than high energies. That is, the *spectrum* of photons, the number of photons with different energies, always has the shape drawn in Fig. 6, a shape characteristic of all ideal radiators. The numerical value of the energy at which the photon spectrum peaks, which is the energy at which the largest number of photons are emitted each second, varies in direct proportion to the *temperature* of the matter in the ideal radiator. Thus an ideal radiator that has a temperature of 6000° above absolute zero will emit a spectrum of photons with a peak at a photon energy of 0.8×10^{-12} erg. An ideal radiator with a temperature of 3000° absolute will emit a photon spectrum with a peak at an energy that is half this much, or 0.4×10^{-12} erg.

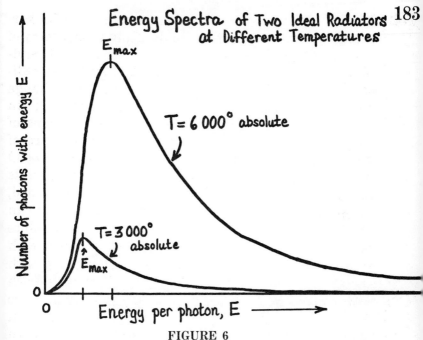

FIGURE 6

The energy spectrum of an ideal radiator has a characteristic shape, no matter what the temperature of the ideal radiator. The ideal radiator will emit the maximum number of photons per second at some energy E_{max}, and E_{max} varies in proportion to the temperature of the ideal radiator. If we double the temperature, the ideal radiator will emit many more photons per second, and E_{max} will double, but the overall shape of the photons' energy spectrum will stay the same. The *total* energy emitted by an ideal radiator at all photon energies varies as the fourth power of the temperature.

Stars like our sun are not ideal radiators. However, the energy spectrum of the photons from a star such as the sun resembles that which an ideal radiator emits, except for numerous dips and wiggles in the photons' energy distribution (Fig. 7). The spectrum of the photons from the sun's surface has a general resemblance to the spectrum of photons emitted by an ideal radiator at a temperature of 6000° absolute. The spectra of the photons

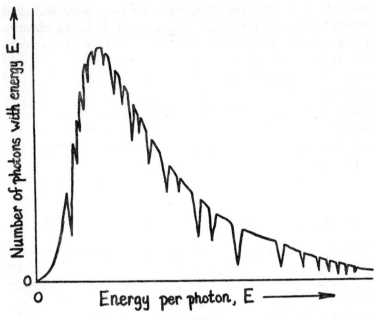

FIGURE 7

A schematic representation of the energy spectrum of the photons
emitted at the sun's surface. The energy distribution of these
photons is similar to that of the photons emitted by an ideal radiator
at a temperature of 6000° absolute, *except* that there are many dips
and wiggles in the spectrum of the sun's emitted photons. These
dips and wiggles occur because certain types of atoms in the sun's
surface layers absorb photons with definite energies and thus
remove these photons from the light that reaches the earth.

from other stars resemble ideal-radiator spectra for temperatures between 2000 and 50,000° above absolute zero.[1]

The dips and wiggles in the spectrum are the result of
the absorption of photons by atoms and molecules in the
cooler, outermost layers of the stars. These atoms and

---

[1]The reason for this resemblance is that for each cubic centimeter inside the
sun or another star there is only a small difference in the amount of energy
emitted and the amount absorbed each second. This difference is highly
important to us because it produces all the sunlight and starlight that we see!
The photons that emerge from the sun's surface have therefore acquired almost
the energy spectrum that characterizes an ideal radiator.

molecules absorb photons with definite energies, and such absorptions produce the jagged spectrum we observe if we separate sunlight or starlight into its various colors (energies or frequencies). By measuring all the dips and wiggles accurately, we can determine which atoms and molecules are present in the stars' outer layers to produce their absorption of certain photons with characteristic energies. In this way, more than 70 different elements have been identified in the atmospheres of our sun and of other, far more distant stars.

SUMMARY

Matter in the form of large aggregates of particles occurs in three possible states: solid, liquid, or gaseous. Its temperature is a measure of the average energy of each of its constituent particles. Matter in any form emits and absorbs photons. If the matter is in equilibrium with its surroundings, it emits the same amount of energy each second as it absorbs and has the same temperature as its surroundings. Matter at a particular temperature emits (and absorbs, if it is in equilibrium with its surroundings) a characteristic number of photons at each frequency. The spectrum (number of photons) that is emitted has a maximum for a photon energy which is proportional to the temperature of the matter. All stars have photon energy (or frequency) spectra that are characterized by the stars' temperatures and the kinds of atoms in their outer layers. These atoms absorb photons of a particular energy (frequency) and modify a star's photon spectrum in distinct ways.

QUESTIONS

1 What is the difference between a liquid and a gas?
2 Does the temperature in a piece of matter increase or decrease if the average energy per particle increases?

3 Does the pressure of a gas in a container increase or decrease when the gas temperature rises? Why?

4 How does an ideal radiator exchange energy with its surroundings?

5 Does an ideal radiator radiate more energy per second than it receives from its surroundings?

6 Which emits more photons per second, an ideal radiator at a temperature of 40,000° above absolute zero or the same ideal radiator at a temperature of 10,000° above absolute zero? How do the values of E_{max}, the energy at which the maximum number of photons are radiated each second, compare for the two cases? (*Ans.*: four times as large)

7 The relationship between pressure and temperature for an ideal gas in a closed container of volume V is $pV = RT$, where p is the pressure, R is a constant, and T is the temperature. How does the pressure change if we double both the volume and the temperature? If we halve the volume and quadruple the temperature? (*Ans.*: p is the same; p increases eight times)

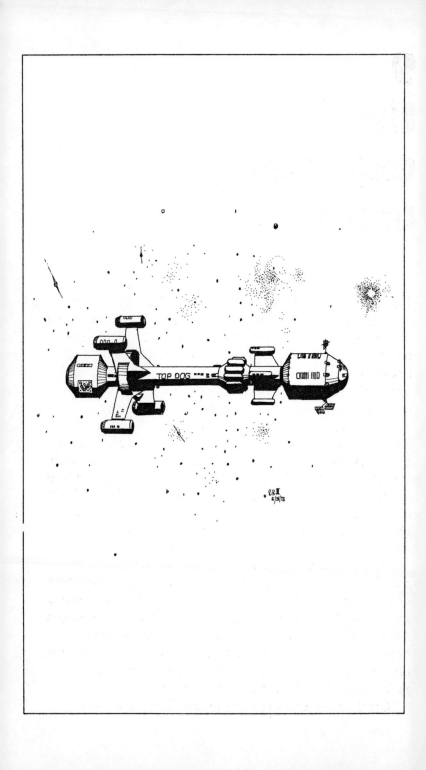

9
the
expanding
universe

The Top Dog leapt from its launch pad with a mighty, fiery, and fearsome bound, thrust outward into space, and began its acceleration toward a velocity of ninety-nine point ninety-nine one-hundredths of the speed of light. Inside, the crew snuggled into their crash couches and began the suid cycle that would keep them from total collapse under the impact of a ten-G acceleration sustained for three weeks. Each man and woman on board the ship felt a tingle mixed of delight and regret in contemplating the series of mind-altering visions that lay ahead. The time dilation that turned years of terminal time into a few weeks on board the Dog left them robbed in their psychic innards, no matter how many times they assured each other that they had gained at the expense of their former contemporaries.

Although Zenith was the most elated passenger, her position as I.E. kept her from showing the eager tremors of anticipation that welled within her like flowers opening toward summer rain. She had never learned what exactly happened to a space crew during acceleration and deceleration. It had been sufficient to know they survived, although a small proportion of the psychorees never pulled out. Direct experience now lay at hand.

"Cyril," Borg enquired of Captain Zaki, who was

doubling himself into a fine imitation of a prenatal child, "what's dilation really like?"

The question came as no surprise, but Zaki's answer suffered from the lack of clarity inherent in the sharing of experience.

"Zenith, I can tell you this. The first time is like no other."

"But what happens, you unfeeling jerk? Does it hurt? Do you thrash around a lot? How big is the risk?"

Cyril remained calm. The suids were doing a fine job.

"Just relax now, Zenith, relax. After a while the visions will take over. You may do some strange things, but you'll feel all warm and open. It happens to everyone. You'll act out things you didn't know were inside you. And if you need someone to depend on, I'll be right here on the A Couch. And I'll try to keep from spacing out completely."

"Do you mean that I'll be saying things I don't want to? Private policies? And doing strange things—what sort of things? Do you mean gross things? Where's my power? What's happening?"

"That's just the suids, Zenith, the keys to cosmic consciousness, as Boyer calls them. They help us get from here to there."

Zenith started to ask another series of questions, but her mind suddenly shifted from her control, and the thought that was entering her brain fell away down a warm smooth tunnel. Cyril Zaki had reentered his childhood and was playing in his crib with his rocket while his ma and pa and sister Vibeke looked on admiringly. And all the while, the Top Dog thrust onward at ever-increasing speed along the preset course toward a point fifteen light years away, near a waning G-type star around which orbited a planet called Thurd. The vast panoply of stars and galaxies that spread across the ship's viewports began to distort, slightly at first, the more dramatically, until with the ship near the speed o

light the entire universe appeared to be concentrated into
a doughnut dead ahead. But by that time the crew paid
no heed because the expansion of the universe meant
nothing compared with the expansion inside their minds.

Many people know that the universe is expanding, but few of us know what to do about it. Some people may ignore the expansion of the entire universe because they do not feel that their own consciousness is expanding. Others, who are tuned in to the universal expansion, may wonder whether or not the universe might someday cease to expand and eventually collapse into a small ball.

Astronomers found that the universe is expanding by measuring the motions of galaxies. Galaxies are huge assemblages of billions of stars, mixed with some gas and dust. What do we mean when we say that the entire universe is expanding? The answer is simply: For distances as large as those between groups of galaxies, *every point in the universe is moving away from every other point, with a speed that is proportional to the distance between the two points.*

We can make a model for the expanding universe by putting dots on a balloon to represent galaxies in space. As we blow up the balloon, each dot on the balloon's surface moves away from every other dot, and the speed at which two dots move away from each other varies in proportion to the distance between them (Fig. 1). The balloon model differs from the universe because the surface of a balloon has only two dimensions, whereas space in the universe has three dimensions.

Another model of the expanding universe is the bumble-bee one. Imagine a huge swarm of bees, where each bee represents a cluster of galaxies. The bees can all obey a "universal expansion" in the following way: Each bee moves away from every other bee with a speed that is proportional to the separation between the two bees (Fig. 2). Then *every* bee will observe that *all the other bees* are moving away from him with a speed proportional to their distance from him. Figure 3 is a diagram of this relationship for the galaxies in the expanding universe.

We can make still another model for the expanding universe by imagining galaxies to be spectators at a sort of cosmic graduation ceremony. Each galaxy rests in a

FIGURE 1
If dots on a balloon represent clusters of galaxies, when
we blow up the balloon we have a model for the expanding
universe. Each dot moves away from every other dot, with a speed
that is proportional to the distance between the dots.

chair separated from its neighbors on either side by 2 ft,
and the rows of chairs are also spaced 2 ft apart. Now
imagine that the chairs actually slide apart from each
other, so that in 1 min the space between each pair of
galaxies has increased to 4 ft (Fig. 4). Then *each* galaxy
would see all the others moving away. At the end of a
minute, the distance to the nearest galaxy would have
increased from 2 to 4 ft (thus at a rate of 2 ft/min), and
the distance to the next nearest galaxy would have increased from 4 to 8 feet (that is, at a rate of 4 ft/min).
Each galaxy would observe that the rate of increase in
the distance to another galaxy varies directly as the
galaxy's distance. This model for the universal expansion has only two rather than the three dimensions of
space, but it can be used to remind ourselves that *all*
galaxies see the same effect as the entire universe expands and that galaxies which are farther away from us
move away faster than the galaxies nearer to us.

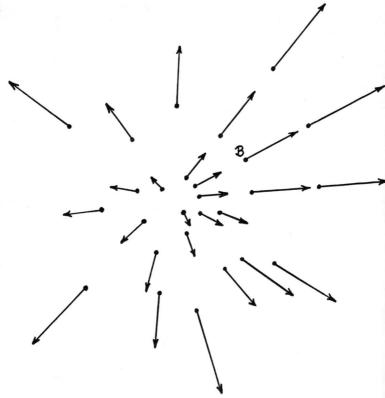

FIGURE 2

An expanding swarm of bumblebees, where each bee represents
a galaxy cluster. The speed of expansion away from the center is
proportional to the distance of each bee from the center. We have
drawn the length of the arrows in proportion to the speed of each bee.
Note that *any* bee, such as the one marked B, will observe *all*
the other bees to be moving away from him, and their speed
of recession from him is proportional to their distance away
from him. (This is similar to the expansion of the universe,
except that the universe had no center because there is no way
to call a given point *the* center of the expansion.)

Each of the three models for the universe has a flaw.
As mentioned, the balloon model has only two dimensions, instead of three as in real space. The bumblebee
model is three-dimensional, but an expanding swarm of

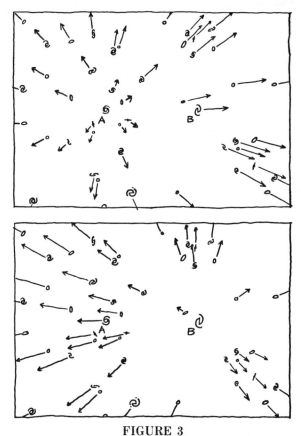

FIGURE 3
Note that no galaxy can consider itself to be *the* center
of the universe just because it observes all the other galaxies
expanding away from it. Both galaxies A and B (and all
the other galaxies as well) observe all the other galaxies
to be expanding away from them.

bees seems to have an outside and to expand from a
definite center. The graduation model has no center, but
it is two-dimensional. In the real expanding universe,
each galaxy sees all the other galaxies moving away
from it, just as the chairs at the graduation ceremony
all move away from one another. No one galaxy at the
graduation can consider itself *the* center of expansion

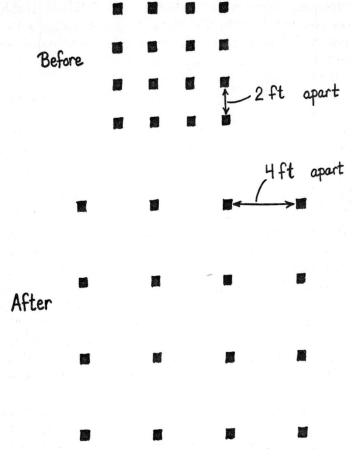

FIGURE 4
We can picture a model for the expanding universe made of chairs
at a graduation ceremony. At first the chairs are spaced 2 ft apart,
but after only 1 min they are spaced 4 ft apart. Then any chair
would observe all the other chairs to be moving away, with
a speed proportional to the distance between the two chairs.

any more than any other galaxy can. In an analogous
experience, *we* can not conclude that the universe has a
center just because we see galaxies moving away from
us in all directions.

How did human beings ever find out that the whole
universe is expanding? In the 1920s, while Babe Ruth
hit home runs and Prohibition kept liquor out of the

stores, an American astronomer named <u>Edwin Hubble</u>
was able to relate the distances of other galaxies to their motions in space. Hubble's measurements revealed that almost all the other galaxies are moving away from our own Milky Way galaxy, with a speed that is proportional to the other galaxies' distances from us. His discovery, first announced in 1929, is now often called Hubble's law. (Astronomers consider Hubble's discovery one of the great achievements of scientific efforts.) Like most "laws" of natural science, it summarizes observed facts in a coherent way.

One example of Hubble's law is the Whirlpool galaxy (Fig. 5), a famous group of stars and diffuse gas whose spiral form almost duplicates that of our own Milky Way galaxy. The Whirlpool galaxy is about 20 million light years away from us; this distance is about 200 times the diameter of our own galaxy, which is 100,000 light years. Measurements of the Doppler effect of the light from the Whirlpool galaxy show that the galaxy is moving farther away from us at a speed of 240 miles/sec. Astronomers have found similar motions that conform to Hubble's law for other famous galaxies and for thousands of other, less well-known galaxies.[1]

Because we believe that the Milky Way galaxy occupies a typical region of space, we can conclude that *all* the galaxies in the universe are moving away from all the other galaxies, in the way that we have described. Of course we must ask: "Do we really occupy an average region of the universe?" We can never know the complete answer to this question because it is always possible that we are in a special position and do not know it. For

---

[1]Only for a very few galaxies do we find motions toward us. These motions reflect the fact that each galaxy has a relatively small random velocity of its own, which is superimposed on the universal expansion. This random velocity can subtract from, or add to, the velocity that arises from the expansion of the universe. Since the expansion velocity increases in proportion to the galaxy's distance from us, only for the nearest galaxies can the random motion be as large or larger than the expansion velocity; thus, only the nearest galaxies are sometimes found to be moving toward rather than away from us.

FIGURE 5

The Whirlpool galaxy, a typical giant spiral galaxy, is made of
billions of stars mixed with some diffuse gas and dust. When we look
at this galaxy we see it almost face on. The Whirlpool galaxy is about

example, our corner of the universe—all the stars and
galaxies that we can see—might be expanding while
other parts of the universe are contracting. However,
astronomers favor the idea that the galaxies that we can
see *do* in fact represent an average sample of all galaxies
in their distribution in space and in their motions. They
have two reasons for this conclusion. First, the galaxies
that we see appear to be uniformly distributed in all
directions. Thus, on the average, the visible universe is
homogeneous (uniform in space). Secondly, we would
have to make an additional, egocentric assumption to
conclude that the part of the universe we can see is
special and different from the universe as a whole. Ever
since Copernicus did his best to convince people that the
earth is not the center of the universe, scientists have
been reluctant to give special treatment to the earth's
position in space.

Our position in space is average in several ways. Not
only are the neighboring galaxies distributed around us
in a uniform way, so that space appears the same (on
the average), but our own sun appears to be an average
sort of star and our own galaxy an average sort of
galaxy. Thus our philosophical inclination to be average,
together with our observation of the uniform distribution
of galaxies around us, leads us to the conclusion that
what we see forms an average sample of what there is.

We should emphasize that when we speak of the universe, by the rules of language we mean *everything that
exists.* The observed "universal expansion" refers, of
course, to everything that we can *see*, but if what we see
is an average slice of the universe, then *everything* really
is expanding away from everything else.

The universe—absolutely everything there is—can be
finite, or it can be infinite. Both possibilities boggle our
imaginations. Who can imagine infinity? On the other
hand, if the universe is finite, where does it end? And
how can it end? We discuss these alternatives in the
next chapter.

Let us first describe how we can measure the *distances* and *motions* of other galaxies. Many years of observations by astronomers have shown that stars are not distributed throughout space evenly. Instead, almost all stars are grouped into galaxies. These galaxies are separated from each other by distances a good deal larger than the average size of a galaxy, and very few stars have been found in the almost empty spaces between the galaxies. When we take photographs through a telescope, other galaxies appear as fuzzy patches of light; the nearest and largest of them, like the Whirlpool galaxy (Fig. 3), can be seen in glorious structural detail.

Because the sun lies rather close to the edge of the Milky Way galaxy (Fig. 6), we can see the especially bright central regions of our galaxy as a bright band or milky way in the summer sky. Our location near the galaxy's edge also allows us to have a relatively clear view out from the galaxy, so that we can see millions of other galaxies. If the earth happened to be in orbit around a star at the center of the galaxy, there would be so much diffuse gas and dust, and so many bright stars around us, that our vision would be highly obscured. In fact, we

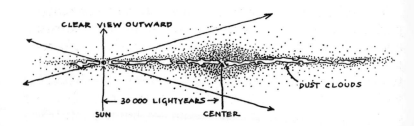

SIDE VIEW OF OUR GALAXY

FIGURE 6

A schematic drawing of a side view of our Milky Way galaxy, showing the sun's location near the outer rim of the galaxy. The diameter of the Milky Way galaxy is about 100,000 light years, but it is only about 5,000 light years thick. Thus we can see outside our galaxy much more easily when we observe in directions perpendicular to the central plane of the galaxy than when we observe along the galactic plane.

would not be able to see any of the other galaxies outside
the Milky Way.

Luckily, we are in a favorable position to observe other galaxies, and with the largest telescopes we can even see the brightest individual stars in some of the nearby galaxies. Within our own galaxy, we can find the distance to a star by making careful observations of the apparent displacement of the star's location in the background of more distant stars and galaxies. This small apparent displacement arises from the earth's motion around the sun (Fig. 7), and it is the same effect that appears if you close first one eye and then the other while studying nearby objects in relation to more distant ones. The earth's orbit is enormous by human standards (186 million miles in diameter) but miniscule in comparison to the distances to stars since the star nearest the sun is about 4 light years, or 24 *trillion* miles, away. Therefore, the earth's motion around the sun causes only a tiny shift in a star's apparent position. This means that the method of determining distances by observing the apparent displacement will work only for the nearest stars because for more distant stars the displacement is too small to be measured.

We can find the distances to very faraway stars by comparing them with nearby stars whose radiation properties and distances we already know. We can make this comparison because two stars that have the same size and temperature emit about the same number of photons, with about the same frequencies, every second. If two stars emit the same number of photons of a given frequency per second, then the two stars have the same *true brightness*. Suppose that we observe two stars with the same size and temperature, but one star appears brighter than the other. We can conclude that the fainter star is farther away from us than the brighter star. In fact, for stars with the same size and temperature, and thus with the same true brightness, the ratio of the stars' apparent brightnesses varies as the square of the inverse ratio of

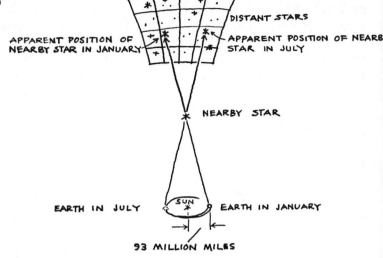

STELLAR PARALLAX

FIGURE 7

The earth's motion in its orbit around the sun produces an *apparent displacement* in the position of a nearby star against the background of more distant stars. If we measure the apparent position of a nearby star at two times that are 6 months apart (say in January and July), then the earth will be at opposite parts of its orbit when we make the two measurements. The apparent position of the nearby star on the background of more distant stars will change by a small angle a from one measurement to another. The distance to the nearby star d can be found from the trigonometric formula

$$d = (93 \text{ million miles}) \times \frac{1}{\sin a/2}$$

The angle $a/2$ is called the star's parallax. The nearer a star is to us, the bigger its parallax because the angle a is larger.

the stars' distances. If we know that a nearby star (whose distance is already determined) and a more distant star have the same size and temperature, we can determine the distance to the faraway star by comparing the two stars' apparent brightnesses.

By analyzing the amounts of light of different colors (that is, of different frequencies) that reach us from various stars, we can determine whether or not two stars do have the same size and temperature, and thus the same true brightness. If two stars emit almost the same *relative* amounts of light at different frequencies, then the stars have the same size and the same temperature. For two stars with the same size and temperature, we might find, for example, that they each emit twice as much blue light as yellow light and three times as much green light as red light. The measurement of stars' spectra also allows us to determine the motion of stars toward or away from us by means of the Doppler effect because if we see a spectrum with ratios that look very familiar except that all the frequencies are changed by a given proportion, we can conclude that we do have a familiar type of star but that it is moving relative to us.

The method of comparing stars of the same true brightness allows us to find the distances to stars that are too far away for the method of apparent displacement to work. By means of this method, astronomers have found the distances to many bright stars in our own galaxy. We can then compare these stars with bright stars located in other galaxies, and if these stars have the same true brightness, we can use the comparison to obtain the distance to stars in other galaxies. Once we know the distance to another galaxy, we can find the total true brightness of that galaxy because we know both its apparent brightness and its distance. We can also compare one galaxy with another, just as we compare one star with another. In this way we can find the distances to faraway galaxies by comparing them with relatively nearby galaxies whose distance is already known (from the bright stars in them) and whose general form and appearance are similar to those of the faraway galaxy.

The nearest big galaxy to the Milky Way is the Andromeda galaxy, which is about 2 million light years (12 billion billion miles away) in the direction of the constellation Andromeda. Galaxies as far away as 5 *billion* light years have been seen, so the Andromeda galaxy is really a next-door neighbor in terms of the visible universe.

To determine the *motions* of distant galaxies we use the Doppler effect. We can spread the light from distant galaxies into its various colors, or frequencies, by means of a spectrograph and then compare this light with the light from familiar elements that can be studied in laboratories, in the sun, or in nearby stars. For example, hot oxygen gas radiates green light at a definite frequency or wavelength. When we observe faraway galaxies, we often find that the galaxies are producing radiation that looks familiar, but the light from the gas in the galaxies has a frequency a little *less* than that which we measure for hot oxygen gas here on earth. Recall that the frequency of light is inversely proportional to the wavelength. The light emitted by distant galaxies has a slightly longer wavelength, and hence is a little *redder*, than is normal on earth. We can conclude that this "red shift" of the galaxies' light to longer wavelengths arises from the galaxies' motion away from us, which is an example of the Doppler effect discussed in the previous chapter.

Hubble spent many nights making careful observations to relate the distances to galaxies to their red shifts (their velocities away from us). Hubble's law, the summary of his observations, says that a galaxy's red shift varies in proportion to that galaxy's distance from us. Once Hubble's law was firmly established by repeated observations, it could be used in the reverse sense. That is, we can find the approximate distance to a galaxy by measuring the velocity at which a galaxy is moving away from us. In this way, we use the galaxy's velocity away from us to place the galaxy in its proper niche in the expanding universe. Thus Hubble's law can be turned around and used to find the distance to strange new objects that have no recognizable stars in them—the quasars.[1] When we use Hubble's law to calculate the distance to a very distant object by measuring its speed

---

[1] We shall discuss quasars and their possible explanation in Chap. 12.

of recession from us and applying the law, we are implicitly making the assumption that we do occupy an average place in the universe and that the universe as a whole is expanding.

For the past 50 years astronomers have had to assimilate the idea that the entire universe, the very fabric of space itself, is changing its "size." Scientists have grown used to the universal expansion concept by using plenty of mathematics and by constantly exposing themselves to the idea. If you find yourself confused by what it means to say that *all* of space is actually expanding, you are in good company. If you like, you can focus your attention on the galaxies all moving away from each other (everywhere!), light up a melloroon, and let the universe spread its wings on its way toward greater size.

We can compare the apparent brightnesses and distances of various objects by using the brightness formula. Recall that if two objects have the same true brightness, then the ratio of the objects' apparent brightness equals the square of the inverse ratio of their distances from us. Mathematically speaking, if two objects called 1 and 2 have the same *true* brightness, then their apparent brightnesses B_1 and B_2 will be related to their distances from us, d_1 and d_2, by the brightness formula

$$\frac{B_1}{B_2} = \left(\frac{d_2}{d_1}\right)^2$$

because the apparent brightness of an object decreases with the *square* of the distance from the object to us.

In our own galaxy, the star nearest the sun, Alpha Centauri, has about the same true brightness as the sun does. Because Alpha Centauri is about 4 light years (24 trillion miles) from the earth, its distance from us is

260,000 times greater than the sun's distance (merely 93 million miles) from the earth. This means that seen from the earth, the apparent brightness of Alpha Centauri (A) compares with the apparent brightness of the sun (S) by the ratio

$$\frac{A}{S} = \left(\frac{\text{distance of sun from earth}}{\text{distance of Alpha Centauri}} \right)^2$$

$$= \left(\frac{93 \text{ million miles}}{24 \text{ trillion miles}} \right)^2$$

$$= \left(\frac{9.3 \times 10^7 \text{ miles}}{2.4 \times 10^{13} \text{ miles}} \right)^2$$

$$= \left(\frac{1}{2.6 \times 10^5} \right)^2$$

$$= \left(\frac{1}{260,000} \right)^2$$

The amount by which Alpha Centauri's apparent brightness is less than the sun's apparent brightness is the square of 260,000, which is about 70 billion (7×10^{10}) times. This makes Alpha Centauri appear to be just another star, although it is one of the 10 brightest stars (in *apparent* brightness) in the sky.

What if we want to compare two objects that do not have the same true brightness? We must then divide the object with the greater true brightness into several objects that *do* have the same true brightness as the fainter object to use our brightness formula. Suppose we want to compare the apparent brightness of the entire Andromeda galaxy with the apparent brightness of the single star Alpha Centauri. To make this comparison, we can consider the fact that the Andromeda galaxy is made of about 100 billion (10^{11}) stars. Let us assume for the moment that *each* of these stars has the same true brightness as Alpha Centauri.[1] The distance to the Andromeda

---

[1] In fact, the average star in the Andromeda galaxy, or in our own Milky Way galaxy, has about one-fourth the true brightness of Alpha Centauri.

galaxy is 2 million light years (1.2×10^{19} miles), which
is 500,000 times the distance to Alpha Centauri (4 light
years or 2.4×10^{13} miles). We can use our brightness
formula to find the ratio of the apparent brightness of a
single star in the Andromeda galaxy (SS) to the apparent
brightness of Alpha Centauri:

$$\frac{SS}{A} = \left(\frac{\text{distance to Alpha Centauri}}{\text{distance to Andromeda galaxy}} \right)^2$$

$$= \left(\frac{\text{4 light years}}{\text{2,000,000 light years}} \right)^2$$

$$= \left(\frac{1}{500,000} \right)^2$$

$$= 4 \times 10^{-12}$$

The brightness formula tells us that a single star in the
Andromeda galaxy with the same *true* brightness as
Alpha Centauri has an apparent brightness that is only
4×10^{-12} the apparent brightness of Alpha Centauri. To
obtain the apparent brightness of the entire Andromeda
galaxy, we must multiply this result by the total number
of stars, 10^{11}, in the Andromeda galaxy. Thus the ratio
of the apparent brightness of the entire Andromeda
galaxy (AG) to the apparent brightness of Alpha Centauri
is

$$\frac{AG}{A} = 10^{11} \times \frac{SS}{A}$$

$$= 10^{11} \times (4 \times 10^{-12})$$

$$= 4 \times 10^{-1}$$

$$= 0.4$$

The entire Andromeda galaxy would appear to be 0.4
times as bright as the single nearby star Alpha Centauri,
if each of the 10^{11} stars in the Andromeda galaxy had the
same true brightness as Alpha Centauri. In fact, because
the true brightness of an average star in the Andromeda
galaxy is only one-fourth the true brightness of Alpha

Centauri, the apparent brightness of the Andromeda galaxy is about one-tenth the apparent brightness of Alpha Centauri. Still, on a clear night the Andromeda galaxy can be easily seen as a fuzzy bright area larger than the full moon; we are seeing light that left the galaxy 2 million years ago, when men had barely established themselves on earth as a separate species.

The Whirlpool galaxy (W) has about the same true brightness as the Andromeda galaxy (AG), but it appears about 100 times fainter. We can use the brightness formula to calculate how much farther away the Whirlpool galaxy is. The brightness ratio is

$$\frac{W}{AG} = \left(\frac{\text{distance to Andromeda galaxy}}{\text{distance to Whirlpool galaxy}} \right)^2$$

$$0.01 = \left(\frac{\text{2 million light years}}{\text{distance to Whirlpool galaxy}} \right)^2$$

From this formula we can calculate that the distance to the Whirlpool galaxy is 10 times the distance to the Andromeda galaxy, so the Whirlpool galaxy is 20 million light years away. On a clear night, people with binoculars can see the Whirlpool galaxy glowing with light that is 20 million years old.

The Whirlpool galaxy is moving away from us at a speed of 240 miles/sec. A much more distant galaxy, located in the constellation of Cygnus, is called Cygnus A. This galaxy is moving away from us at a speed of almost 10,000 miles/sec, or 40 times the speed at which the Whirlpool galaxy is moving away from us. Hubble's law then allows us to calculate at once that Cygnus A is 40 times farther from us than the Whirlpool galaxy, or 800 million light years away.

Let us consider the Doppler shift of a star or galaxy, which we discussed in Chap. 7. When a photon is emitted from a source that is moving toward us, the photon appears to be more energetic (it has a higher frequency)

than it would if the source were stationary. Conversely,
when a photon is emitted from a source moving away
from us, the photon has a lower energy (or lower frequency) than it would if the source were stationary (see
Figs. 4 to 6, Chapter. 7). The ratio of the frequency of a
photon emitted from a moving source to the frequency
that the photon would have if the source were stationary
is given by

$$\frac{\text{Observed frequency}}{\text{Frequency for stationary source}} = \sqrt{\frac{1 - (v/c)}{1 + (v/c)}}$$

Here v is the velocity of the moving source, and c is the
speed of light. The velocity v is positive if the source is
moving away from the observer and negative if the source
is approaching the observer.

Consider a star that is moving away from us with a
velocity $v = 0.6c$. The star emits a group of photons that
would have a frequency of 10^{15} cycles/sec and would
appear blue if the star were stationary. We can use the
Doppler formula to calculate the shift in frequency of
the photons that we observe from the star:

$$\frac{\text{Observed frequency}}{10^{15} \text{ cycles/sec}} = \sqrt{\frac{1 - (0.6c/c)}{1 + (0.6c/c)}}$$

$$= \sqrt{\frac{1 - 0.6}{1 + 0.6}}$$

$$= \sqrt{\frac{0.4}{1.6}}$$

$$= \sqrt{\frac{1}{4}}$$

$$= \frac{1}{2}$$

That is, the observed frequency is 0.5×10^{15} cycles/sec,
which is the frequency of red light. The original blue
light has been reddened or red-shifted so much that it

now appears red. Since the frequency f of a photon is inversely proportional to its wavelength (see page 152), the wavelength of these photons has increased by a factor of 2.

SUMMARY

Observations of other galaxies (large clumps of stars) show that all galaxies except some of the nearer ones are moving away from our own Milky Way galaxy. The speed of the galaxies' recession increases in proportion to their distances from us (Hubble's law). If we assume that our galaxy occupies an average region of space, then all galaxies must be moving away from all other galaxies, everywhere. This is taken to mean that the entire universe (all of space and everything in it) is expanding. Such an expansion does *not* mean that the universe has a center.

Hubble's law describes the proportionality between the galaxies' velocities and their distances from us. We measure the velocities toward or away from us through the Doppler effect. To determine the distances to faraway objects like galaxies, we compare their apparent brightness with that of a nearer object whose distance we already know. The apparent brightness of an object decreases as the square of its distance from us. Thus if two objects have the same *true* brightness, the ratio of their apparent brightnesses varies in inverse proportion to the *square* of their distances away from us. We can use this relationship to find the distances to faraway objects if we know their apparent brightness and compare it with the apparent brightness of an object whose distance from us is already known.

QUESTIONS

1   Is the universe expanding or contracting? How do we know?

2 Where is the center of the universe?

3 Do you think that the idea that our position in space is average is a good one?

4 Suppose that we were located so close to the center of our galaxy that we could not see other galaxies because of obscuring clouds of gas and dust all around us. Would we have discovered that the universe is expanding?

5 If we observe two spiral galaxies that seem to have the same general appearance, but one is 100 times greater in apparent brightness than the other, which galaxy is likely to be farther away? By how much? (*Ans.*: 10 times farther away)

6 The bright star Rigel in the constellation Orion has a true brightness about 10,000 (10^4) times greater than that of Alpha Centauri. However, Rigel is about 100 times farther away from us than Alpha Centauri. How do the apparent brightnesses of Rigel and Alpha Centauri compare with one another? (*Ans.*: they are the same)

7 Suppose that we measure the velocities of two very distant galaxies, and we find that one of them is moving away from us at a speed of 5,000 miles/sec while the other is moving away from us at 25,000 miles/sec. Which galaxy is farther away from us? By how much? (*Ans.*: five times farther away)

8 What is the opposite of a red shift? What does this kind of Doppler shift mean?

9 Can the universe expand with nothing to expand into? (Remember that the term universe means *all* of space.)

10 Could the universe contract with nothing to contract into?

10
space, time, and us

Zenith Borg felt the ecstasy of spaceflight for the first time. She found herself expertly manipulating a giant lingam that seemed to grow from the center of humanity, and realized that once more she had the destiny of womankind in her grasp. And at the moment of this realization, she slipped from the bonds that had closed her mind, to begin a new way of thinking, unencumbered by the old directions, numbers, and calculations. There was no more up or down, left or right. An entire raft of mental dross had floated away. The essence lay straight down the river.

Into the feathered recesses of her new-found wisdom swam the familiar presence of Cyril Zaki, rough, hairy, and toothsome.

"Good risk, eh, baby?" he inquired.

"Cyril," answered Zenith with passion, "there are no landmarks in space or time. I see it all now."

"Sure you do, Zenith. In time everything becomes clear."

A sudden whiff of worry came into her vision.

"Don't talk to me about time, you space sweeper. I can go forward or backward in time—what you would call forward or backward. And in space too. There are thousands of ways to travel, not just three dimensions.

You're bound down by that galactic latitude and longitude nonsense. Take a look and see how things really are."

As she spoke, Zenith saw clearly what she meant. A forest of rigid axes appeared, each one perpendicular to all the rest. The sparkling rods waved an invitation to follow first one, then another.

"Cyril," said Zenith at the peak of her being, trying to put all the intensity she felt into a single effort of communication,"give me some assurance. There is no back and forth. No in nor out. In! Out! In! Out!"

Cyril seemed to understand for the first time. His huge hands trembled slightly.

Experience shows us that space has three dimensions. We rarely pause to think about what this means, or whether it means anything. Great thinkers have argued about whether the three dimensions of space arise by accident or by design, but almost all agree that the number *is* three. The theory of relativity implies that although we must consider time in some respects to be another dimension, we must always remember that time and space *are* different for us. In space we can move up or down, left or right, forward or backward, but in time we can proceed in only one direction—onward. We shall return to the time dimension a little later, but for now we shall consider the three dimensions of space.

These three dimensions create the geometry[1] of the universe. Most of us have studied the plane geometry of two dimensions. By limiting ourselves to *two* dimensions in space, we can step back into the third dimension and examine the other two dimensions. We do this when we step back from a blackboard to get a good look at a circle or triangle we have drawn on it.

The plane of two-dimensional plane geometry is the flat surface—a blackboard or a piece of paper—on which we draw figures. Plane geometry deals with two-dimensional, *flat* space. Since there is no curvature of the flat surface inward or outward, we can extend the surface to infinity in our imagination. It is not very hard to visualize a blackboard growing wider and wider, covering a greater and greater area, without any limit. This is also the way that most of us think of the three dimensions of the universe. What else could be going on?

Our own earth gives us the answer because it provides us with a familiar *curved* surface of two dimensions (Fig. 1). Most of us easily accept the fact that the earth's surface is not flat; rather, it is curved, or bent right back onto itself, in an almost spherical shape. This curvature means that if we set out in what we think is a straight

---

[1] In Greek, "geometry" means "measure of the world."

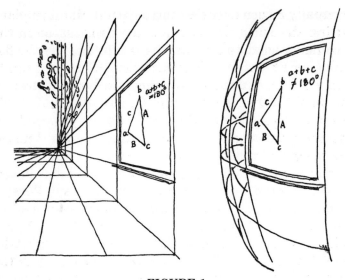

FIGURE 1
An ordinary blackboard is an example of a two-dimensional *flat* surface; the surface of a sphere is a two-dimensional *curved* surface.

line, and continue to travel in the same direction, we will eventually come back to our starting point, no matter what direction we follow. In addition, the surface of the earth is *finite*, whereas a *flat* two-dimensional surface would go on indefinitely. That is, curved surfaces bend back on themselves, and include only a finite area, but flat surfaces are theoretically quite infinite (Fig. 1).

Similarly, the entire universe—all of space—may be flat, or it may be curved. Flat space extends to infinity, but curved space is finite. If space in the real three-dimensional universe is curved, then the universe has a finite volume, just as the curved surface[1] of the earth has a finite area. Furthermore, in a curved universe, continuous travel in what we call a straight line would

---

[1]Strictly speaking, what we have called curved surfaces are "positively curved surfaces" to mathematicians. There are also "negatively curved surfaces," which are infinite, as flat surfaces are; we shall not discuss them here.

eventually return us to our starting point, like a jet plane circling the earth. If, on the other hand, space in the universe is flat, then its volume is infinite, like the flat surface of an infinite blackboard, which has an infinite area. In a flat universe, continued motion in a straight line would never return us to our starting point. This operational test would give us a real way to find out whether space is flat or curved, provided we had a rocket ship to travel in for billions of years.

Because we have an extra, third dimension to think in, we can easily see how the outside of the earth forms a curved, two-dimensional surface. We can move east or west, north or south on the surface of the earth, but motion in the third dimension—up or down—would take us off the surface. When we study the geometry of the entire universe, there is no good way to get a view of three-dimensional space because we can not step back into a fourth dimension. Instead, as prisoners of our three-dimensional universe, we must study and imagine it *from the inside*. When we try to imagine the curvature of all space, of the entire three-dimensional universe, we are cruelly handicapped because we have no extra dimension. No wonder then that we can not picture just how the universe could be finite and still have no boundaries! But this might be reality, even if we can not imagine it accurately. The universe would then be like the two-dimensional curved surface of an expanding balloon, only it would have three dimensions, and we do not have the extra dimension that allows us to understand the expanding balloon model for the universe.

We can easily visualize the curved two-dimensional surface of an expanding balloon precisely because we have a third dimension into which we see the surface curving; thus we can locate the "center" of the expansion *in the extra dimension*. But suppose that we were two-dimensional creatures sliding over a smooth sphere (like a balloon) with no concept of "up" or "down." Even though we could detect the sphere's curvature (for example, by

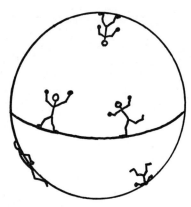

FIGURE 2
If we were two-dimensional creatures slithering over the surface
of a sphere, we would imagine that the entire universe was just
the curved surface we lived on, unless we could discover the
third dimension by lifting our heads to find "up" and "down."

sliding all around it),[1] we would find it almost impossible
to imagine how this curvature might "look" (Fig. 2). This
difficulty really does occur in our three-dimensional space,
which may be flat or curved, because two-dimensional
space can be flat or curved. We can not hope to visualize
the possible curvature of three-dimensional space, but
still the curvature may exist.

We know that the curvature of the earth's surface
means that the surface has a finite area. An earth crea-
ture who had no third dimension would conclude that
his "universe" is finite, and it would be, except that there
is a third dimension (Fig. 3). But three-dimensional
space might also be finite, and yet be everything that
exists because there is no fourth dimension. Couldn't
there really be a fourth dimension? A little reflection
will convince us that a fourth dimension is, to say the
least, highly unlikely. For one thing, a fourth dimension

---

[1]We could make a mark on the sphere's surface, slide along in a "straight"
line, and discover that we had come back to the mark. This would tell us that
the sphere is curved.

FIGURE 3
If we had no third dimension, we would think that
the surface of the earth was the entire universe.

in space would provide a way to reach any point now
hidden in three dimensions. For example, every point
inside our bodies would be accessible from the outside
without touching our skin. No building could contain
us, any more than a circle can contain a boy who knows
how to step over it.

Let us look at the so-called fourth dimension: time.
Because we can label events not only as to where they
occur but when they occur, we can see that in some ways
time is a dimension. However, for us space and time are
not interchangeable. Although modern physics and quantum mechanics have taught us not to be too sure that our
everyday experience about space and time always corresponds to reality, it still is possible to separate space
from time in describing our world. When we describe the
time dimension of the universe, we can divide events

Earlier time

Later time

FIGURE 4
The expansion of the universe has produced a constant decrease
in the average density of matter. Galaxies used to be closer
to one another than they are now, and the further back in the past
we go, the greater the density of matter must have been.

into past and future. This division into past or future
time, so important and obvious to us on earth, remains
a key feature of the time dimension even in relativity
physics.[1]

To reconstruct the past history of the universe, we
recall that the universe is expanding: All galaxies are
moving farther away from one another as time passes.
This means that in the future galaxies will be farther
apart from one another than they are now. In the past,
galaxies were closer together than they are now (Fig. 4).
The farther back in the past we go, the more closely

[1]Relativity theory shows that we can not *always* tell which of two events
occurred before the other. However, we *can* always construct a rational
description of our physical world according to which no event can influence
another event unless it really *did* happen at an earlier time. Thus relativity
theory defines time in a way that avoids the philosophical contradictions of
allowing the future to influence the past. This proper sequence of events in
time guarantees what physicists call "causality" and what we might call
common sense.

packed together galaxies must have been. Since the universe in the past had less volume than it does now, its density (mass per cubic centimeter) must have been greater than it is now.[1] If we continue our mental excursion into the past, we find that at some time long ago all the material now in galaxies was packed to an incredibly high density, with the density increasing higher and higher as we mentally approach the beginning of the universe. Mathematicians use the word "singularity" for this packing point of unlimited density. It seems unlikely that the universe began as a "singular point" with infinite density; instead, the universe probably started to expand from an extremely dense cluster of matter and energy, far denser than anything we can find today.

No one really knows how dense the universe was when it started expanding. For that matter, no one knows how the universe came to exist in the first place. However, we can calculate from the present rate of expansion that the universe began to expand about 15 billion years ago. For comparison, the solar system is about 5 billion years old, and the oldest rocks found on the earth and moon are $4\frac{1}{2}$ billion years old.

What future can we foresee for the universe as a whole? There are basically two possibilities: Either the universe will continue to expand forever, or the expansion may someday end, to be replaced by a universal contraction. An eventual contraction of the universe would produce a return to a cluster of immense density, cancel all our errors and achievements, and give the universe a chance to reemerge in a new cycle of time. On the other hand, the universe may go on expanding for all time. An infinite universe will expand forever, but if the universe is to ever contract, it must be finite. Surprisingly enough, we could discover whether the universe is finite or infinite, and thus learn the ultimate fate of the universe, if

---

[1]We are assuming that the total amount of matter and energy in the universe remained constant as the universe expanded.

we could measure one critical parameter with sufficient accuracy. This parameter is the average density of matter in the universe. The density of matter is the amount of matter (or mass) per unit volume. The total amount of matter in the universe equals the average density of matter times the total volume of the universe. Now, why is the average density so important?

Galaxies are systems of stars in which the stars interact with one another through gravitational forces. The galaxies themselves also interact with other galaxies by gravitational forces between them. The overall characteristics of the motions of galaxies, that is, of the universe's expansion, are determined by the gravitational forces within the universe. The strength of these forces depends on the amount of mass (matter) or, in other words, on the average density of all the material in the universe. We can think of the entire universe as having somehow been given a terrific shove 15 billion years ago that started it expanding.

Whether the universe will expand forever depends on the strength of the gravitational forces among its component parts, and this strength in turn depends on the average density of matter. At each moment in the history of the universal expansion there is one critical value for the average density of matter in the universe. If the actual average density of matter is *larger* than this critical value, gravitational forces will eventually stop the expansion of the universe and cause the universe to pull itself together. On the other hand, if the actual density is *smaller* than the critical value, the universe will always expand.[1] The critical value for the density is now 5×10^{-30} gm/cm^3. This tiny value is equal to the mass of three protons per cubic *meter* and seems incredibly small. Tiny as the critical density is, it determines the future of the universe.

---

[1]Both the actual density and the critical value for the density decrease in the same proportion as the universe expands, so if we can answer the question of which is larger at any one time, we will have the answer for all time.

If the actual density of matter in the universe now exceeds 5×10^{-30} gm/cm^3, then the universe will eventually contract. The universe will then reach only a finite size and in fact must be finite now and forever. On the other hand, if the average density now is less than 5×10^{-30} gm/cm^3, the universe will never stop expanding. This means that even if the universe is finite now, it will grow larger and larger forever and will reach an infinite size in an infinite amount of time but not in any finite time. There is also the possibility, if the actual average density of matter is less than 5×10^{-30} gm/cm^3, that the universe has always been infinite. In the long run, both alternatives for a density less than the critical value are equivalent.

Astronomers have two ways of determining the actual density of matter in the universe: (1) to observe the density of galaxies and assume that this gives us the density of all matter (from a careful study of our own galaxy, we know how much mass there is in an average galaxy); (2) to observe the *motions* of the galaxies in the universal expansion and see how these motions compare with the theoretically calculated motions for different possible densities of matter. We can use the second method profitably only if we can see the universe at several different times because an observation at just one time simply tells us that the universe is now expanding at a certain rate. If we can look at the universe at several different times, then we can compare the rate of expansion at different times and see whether the expansion rate is changing, and by how much.

But we are here only right now; our lifetimes are a flickering instant in the history of the universe. How can we possibly see the universe at several different times?

We can do this because when we observe very distant galaxies, we are looking a long way back into the past. Light from these distant galaxies has taken many millions, even billions, of years to reach us, so we are looking back a certain fraction of the 15-billion-year expansion history, seeing the galaxies and their motions as they were a long time ago. This glimpse of past ages of the universe allows us to compare the motions of galaxies at earlier times with the motions that we see at almost the present time for relatively nearby galaxies. Such a comparison, together with our calculations of how the changes in the galaxies' motions relate to the density of the universe, gives us the value for the density of matter. Both methods for determining the average density of matter in the universe have been used, but they have given somewhat contradictory results.

Observations of the density of galaxies in space have shown that galaxies do not contain enough matter—they provide an average density about 30 times less than the critical density—to cause the universe to ever stop expanding and start contracting. On the other hand, studies of the motions of galaxies suggest that the density of matter in the universe *is* close to the critical density. Both these observational tests involve very difficult measurements, and the results are not yet conclusive. However, we should consider what it would mean if they were later verified.

If the galaxies provide an average density far less than the critical density, but if, nonetheless, the average density in the universe does nearly equal the critical value, then most of the matter in the universe must be in some form, as yet undiscovered by us, that does not shine like stars in galaxies, or we would already have seen the material. At least two forms for this "missing mass" are possible. One would be diffuse gaseous matter (or for that matter, sand and rocks, but that is unlikely) that has not yet formed stars or galaxies. The other possibility is that there may be huge numbers of burned-out stars or galaxies that have exhausted their energy-producing cycles and have ceased shining. Such stars will contract under the influence of their own gravitation because they no longer have the support of the energy liberated in their centers to push outward on the overlying layers. Some of these stars might have shrunk to smaller and smaller sizes with greater and greater densities. Eventually such stars would become invisible to us. They would have only a gravitational interaction with the rest of the universe. In fact, they would be black holes.

SUMMARY

The universe (all of space and everything in it) may be finite, or it may be infinite. A "flat," infinite space would go on and on forever, but a "curved" space is finite. The possibility of a finite universe does *not* mean that the universe ends somewhere; it means that space somehow bends back on itself, so that continuous motion in a straight line would eventually lead back to the starting point, like going around the earth. We can make only imperfect models for the *entire* universe because we are used to stepping back from a model to examine it. Three-dimensional space in a finite, expanding universe would be analogous to the two-dimensional surface of an expanding balloon, without an extra dimension to use in understanding how the two-dimensional curvature "looks."

The universe has been expanding for 15 billion years. We could determine whether or not the universe will expand forever *if* we could measure the average density of matter in the universe. If the average density now is greater than a certain critical value, the universe will eventually contract, but if the average density is less than this critical value, the universe will expand forever. The amount of matter shining in stars and galaxies seems to be about 30 times less than the critical value. However, when we look back in time (by observing faraway galaxies) to see how the expansion speed has changed, we find indications of a density close to the critical value. Perhaps most of the mass in the universe is in some nonshining form of matter, such as black holes.

QUESTIONS

1 How many dimensions does space have? How would you describe them?
2 If the universe had only one dimension, what would it look like? What would people look like then?
3 Have you had any success in imagining a fourth dimension for space?
4 The surface of a balloon is a curved two-dimensional surface. What is an example of a curved one-dimensional surface?
5 How long has the universe been expanding? Is this more or less than the age of the earth?
6 How can we decide whether or not the universe will expand forever?
7 What sort of forces determine whether or not the universe will expand forever?
8 How can we measure the density of matter in the universe?
9 If the universe is infinite, will it ever contract?
10 Infinity is a strange number. For example, both the integers (1,2,3, etcetera) and the even numbers (2,4,6, etcetera) are infinite sets. Do you think that there are more integers than even numbers? Would there be more space in an infinite universe after it had expanded for 15 billion years more?

1 If the universe is infinite, can you see any limit on the number of possible worlds and life forms? For instance, in an infinite universe, is it likely that this book exists in all possible languages? With all possible variants? For all possible life forms of readers thinking all possible thoughts?

2 If the universe starts to collapse and then starts expanding and repeats this cycle an infinite number of times, do you think we would all live our lives over?

11
the
fingerprint
of time

The cabin gradually came back into view, and Zenith
was able to focus her vision on the dials in front of her,
then on the control area, and finally on the bridge with
its black sphere astride. Borg saw the crew talking quietly
and earnestly. An aura of good feeling seemed to
permeate the Top Dog's interior.

"Hi there, your Assurance," said Bernie Zots. "How
does it feel to be a spacehead?"

"Not bad, Bernie," answered Zenith. She wondered
how she knew his name, and whether he wouldn't prefer
to be addressed as Officer Zots.

The massive figure of Captain Zaki arose from
behind the console, where he and Julie Card had been
examining the torquometer.

"Hey there, Zenith, how's the risk? See why space
travel takes a triple premium? Sort of wears you out,
doesn't it? Here, have a zinger."

"I don't smoke, Cyril, you know that," but Zenith
found herself lighting the melloroon Zaki gave her. She
felt far less distance between the crew and herself than
she would have expected from her position as I.E., and
somehow they seemed to share this feeling.

Lisa Dalby was first to pose the key question that
hovered menacingly behind the good waves in the cabin.

"Zenith," she asked, "what's Zed up to? What can he do when we land on Thurd?"

For a moment Borg hesitated. If she admitted that she didn't know all about Zed, the crew might bad-risk her and succumb to fear. But that danger had to be met if they were to overcome the malevolent Bokomoru. Zenith spoke in clear, ringing tones.

"Women and men, I'll write you an honest policy. We've been keeping track of Zed as best we can, but even the Pilar computer couldn't calculate his premium exactly. We think he's found out that there's old knowledge stored in the Permids. Zed is desperately eager to control powerful energy sources. So he planted a sender in Captain Zaki's bubke and tricked us into saying that the Permids are on Thurd. Now he's on his way there to root out what he can."

"But what will Zed do if he gets the oldtime knowledge?" Julie Card asked.

"Well, the data point only one way: He's out to overthrow the Trust itself."

Stifled gasps burst from the taut assembly. There was a flurry of melloroons being lit. Zaki took a deep toke and reflected on his encounter with the wizard of Sidney. What was it that Zed had told him about his discoveries?

"Uh, Zenith," he broke in. "When Zed talked about orgone energy, was that the power source you meant? Because I thought he was just loose in the computer."

"Well, Cyril, it could be. We know that space is filled with radiant energy—microwaves are everywhere, going right through this cabin. You know, Cyprian called the microwave background the 'ultimate everything' and the 'instant forever.' Suppose Zed found a way to pick up on those resonances! He'd be writing the policies then, instead of the Trust!"

Zaki nodded gravely. He clenched and unclenched his huge hands. Imagine the gall of that Zed to think he was capable of replacing Zenith Borg and the Trust!

There was just no end to the bad vibrations some people
were willing to tune on. And what about these micro-
waves—waves Zaki couldn't see—were they really in all
of space? Wouldn't they get in everyone's way then?
Cyril began to mutter to himself.

"I'll teach Zed how to relax. I'll show him a thing or
two, my sweet boogaloo."

Triggered by the keyword "boogaloo," the crew forgot
about the evil Zed and burst into the Dog's fight song,
which commemorated their astonishing success on
Xanaroo:

> *In Xanaroo did Zaki dance,*
> *His boogaloo was waving free,*
> *And all could see his spangled pants,*
> *His brilliant vest, his mighty hands,*
> *Down by his dimpled knees.*

The chanting was followed by more melloroons all
around, and the crew soon produced some broad smiles
and hearty chuckles.

Cyril felt good since it was his job as captain—and
I am the captain, let's make no mistake about that, Zaki
thought—to keep the crew from being bad-risked. He
felt the warm glow of contentment sweep over him.

The expansion of the universe from an extremely dense cluster of matter and energy to its present form has taken about 15 billion years. Most of what we know about the universal expansion and the early history of the universe has come from studying tremendously distant *galaxies*. These galaxies are so far away from us that they appear on photographs as tiny, fuzzy dots after hours of exposure through the world's largest telescopes; they are so far away that the light from them, traveling at a speed of 6 trillion miles/year, started on its way to us long before man appeared on this planet.

Until recently, all our information about the universal expansion had been obtained by observing galaxies. But since 1965 we have gathered further information about the early history of the expanding universe by studying *microwave photons* that fill the entire universe, including the regions around the earth. These microwaves were produced during the earliest years of the universe and now spread through all of space as a universal flux of radiation. This universal flux forms a background to all the other radio waves that we can detect. The flux is often called the *universal microwave background* because most of the photons that compose the background flux have energies typical of the short radio waves called *microwaves.*[1]

To understand how these microwave photons came to permeate the entire universe, we must consider the universe as it was in its early moments, just after the start of its expansion. At this point, the universe was an extremely dense conglomeration of matter and energy, often referred to as the "primeval fireball." Because the primeval fireball had such a high density of matter and energy, it also had a very high temperature. The "known" history of the universe starts with the expansion of this incredibly dense and energetic primeval fireball. During the first few seconds of the fireball's existence, particles

---

[1]Figure 3, Chap. 7 (page 153) shows the frequencies and wavelengths of microwave photons.

were so densely packed together that all sorts of strong and weak nuclear reactions occurred among them at a fantastically rapid rate. Every type of particle and antiparticle was created from other particles in vast profusion, and these particles decayed into other particles and annihilated one another, thus producing energetic photons, neutrinos, and antineutrinos. Out of this universal roar of nuclear reactions came the "ordinary" particles that we see today. After the first minutes of the universal expansion, the fireball had expanded (particles were farther apart), and its density had decreased to the point where strong and weak nuclear reactions no longer occurred rapidly enough to change the composition of the universe in a significant way.

We do not know what elementary particles were present when the fireball started expanding, but after the first few minutes of the expanding universe the basic mixture of particles had emerged from the rush of particle collisions. We think that this basic mixture is primarily made of protons, neutrons, helium nuclei, and electrons, plus photons, neutrinos, and antineutrinos. Our own galaxy consists mainly of these particles, and we extrapolate from our notion that our own galaxy is average. The total electric charge of the universe seems to be almost zero; that is, the total negative charge of the electrons came out almost equal to the total positive charge in the nuclei. (Recall that photons, neutrons, neutrinos, and antineutrinos have no electric charge.) There is no good explanation of why this balance of positive and negative electric charges exists.

The protons, neutrons, and electrons that now form our familiar atoms must represent the balance of particles (matter) relative to antiparticles (antimatter) which remained after a huge number of annihilations between particles and antiparticles in the first few minutes of the universal expansion. However, it is possible that entire other galaxies might be composed of antimatter (antiprotons, antineutrons, and positrons) rather than ordinary matter. Such galaxies and their component stars would behave just like our own and would produce photons in the same way as ordinary stars and galaxies do, but if our galaxy were to collide with an antimatter galaxy made of antiparticles, a titanic mutual annihilation would blow the galaxies apart. If antimatter galaxies do exist, they must be far removed from the galaxies made of ordinary matter, or we would expect to see an occasional spectacular annihilation, which has not yet been observed. On the other hand, it may be true that *all* galaxies are made of ordinary particles and that the preponderance of particles over antiparticles, which is overwhelming in our galaxy, also exists throughout the universe. This question is still unanswered.

We shall discuss the universe as if it were made predominantly of ordinary matter. Our conclusions about the universal background flux of microwave photons would not be changed if it were found that half the universe is made of antimatter. The reason for this lack of change is that photons (electromagnetic radiation) interact with matter and antimatter in the same way because photons and antiphotons are identical.

To understand the universal microwave background, we must consider the primeval fireball or "big bang" that began the universe in a little more detail. The big bang had so many nuclear reactions in its early moments that fully 25 percent of the hadrons (protons and neutrons) were fused into helium nuclei, each with two protons and two neutrons. After the first few seconds of the primeval fireball, about 25 percent of the mass of the universe was helium nuclei, and almost all the remaining 75 percent was individual protons. There were very few isolated neutrons (remember that these tend to decay), and although there were large numbers of electrons, neutrinos, antineutrinos, and photons, they totaled only a tiny fraction of the total mass since an electron's mass is 1/1,836th of a proton's. All the nuclei that are heavier than helium, such as carbon, oxygen, or iron, were produced only in tiny fractions (less than one-millionth of the number of protons) during the first seconds of the big bang.

After the universe had expanded for its first few minutes, the density and temperature of the matter throughout the universe were no longer high enough for strong and weak nuclear reactions to continue in the universe as a whole. At this point, the universe must have consisted almost entirely of hydrogen and helium nuclei, plus electrons *that were not in orbits around the nuclei*, neutrinos, antineutrinos, and a great number of photons. The matter in the universe was constantly creating and absorbing photons in an effort to reach equilibrium with its surroundings—which in this case was the universe itself. All these particles were spread out *diffusely* in space. The clumping process that formed stars and galaxies did not begin until the universe had been expanding for millions of years. The photons filled the universe with electromagnetic radiation, and they continue to do so today.

A strange thing happened, though, after about 300,000 years, long after the expansion started but long before any matter could clump into stars or galaxies. By this time the average energy of each photon in the universe, which had been produced by the particle collisions and annihilations of the first few minutes, had decreased to about half a trillionth (0.5×10^{-12}) of an erg. The energy of a photon is proportional to its frequency and inversely proportional to its wavelength. Now, the wavelength is a distance, and the universal expansion makes all distances grow larger. That is, *particles* do not expand, but distances do, which means that as the photons traveled through space, their frequency or energy slowly *decreased* because their wavelength *increased* as the universe expanded. The photons that were produced at an earlier state of the universe have less energy now than they did when they were produced and radiated away from particle collisions. (If we prefer, we can view the increase in the photons' wavelengths simply as the Doppler shift. At any moment in time, photons created during earlier times and reaching us now have red shifts characteristic of their moment of creation. Because the entire universe is expanding, all the photons from earlier times are red-shifted since they come from a place that is receding from us: the entire universe.)

The critical point in the photons' history came after 300,000 years because the photons no longer had enough energy to *keep atoms from forming*. Until then, almost no electrons could combine into lasting atoms by orbiting around a proton or a helium nucleus. If an atom did form in this way, it was almost immediately destroyed when a photon hit the atom and used its energy of motion to knock the electron apart from the nucleus. But when the average energy of a photon had decreased to half a trillionth of an erg, almost none of the photons had enough energy to break atoms apart in this way. About 300,000 A.B.E. (After the Beginning of the Expansion), the particles in the universe combined into atoms. Then, the photons traveled between the atoms without breaking them apart because the photons had too little energy to do so. Since 300,000 A.B.E., which was not long after the big bang (by our standards), the photons and the particles with mass have essentially gone their separate ways. The microwave background radiation comes to us from the moment billions of years ago when atoms were first formed, leaving the photons to travel through space without any significant interaction between themselves and the atoms. All that has happened to the photons since then is that the expansion of the universe has increased all their wavelengths (and all distances) about 1,000 times since 300,000 A.B.E., which has caused the energies of the microwave photons to decrease by a factor of 1,000.

The photons in the universal microwave background, left from the early years of the universe's expansion, are the oldest observed "fossils" in the universe. These photons have not interacted significantly with any matter since 300,000 A.B.E., about 15 billion years ago. They preserve a record of what the universe was like then, a record that we can interpret once we allow for the fact that the universe has expanded a thousandfold since

then. When we consider that these cosmic relics come
to us from a time when the universe was 1/50,000th of its
present age, we can see that we now know how to listen
to the cosmic OM in the murmur of time.

One additional level remains to be reached as we
search for the early relics of the universe: the neutrinos.
All the universe should be filled with neutrinos and
antineutrinos that were produced during the first few
minutes of the primeval fireball. These neutrinos and
antineutrinos have not interacted with anything in a
significant way since the first hour after the big bang.
Unfortunately, neutrinos and antineutrinos are almost
impossible to detect because they pass through ordinary
matter with almost no interaction. This is why the cosmic flux of neutrinos has remained unchanged, except
for a steady decrease in the energy of each neutrino
(arising from the universal expansion) since a few minutes A.B.E. On the earth, the greatest flux of neutrinos
comes from the interior of the sun, where neutrinos are
being made during the first step of the proton-proton
cycle that liberates energy in the sun (we described this
reaction on page 71). These solar-made neutrinos pass
right through 400,000 miles of solar material by the
trillion-trillion-trillions each second, and the detectors
that we have now manage to catch less than one each
day. The solar neutrinos overshadow the universal neutrino background by thousands of times. Still, someday
we may be able to find the flux of universal neutrinos and
thus be able to tune in not merely to the first 300,000
years of the universe but to its first hour!

★　★　★　★　★　★　★　★　★　★

The universal background of microwave radiation has a
characteristic *energy spectrum*. The spectrum of a group
of photons gives the number of photons with different

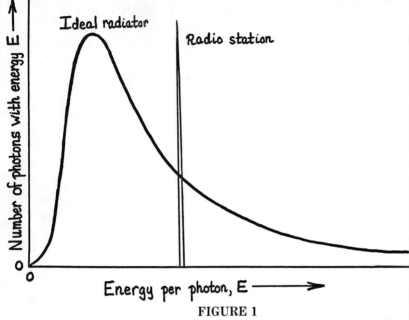

FIGURE 1
A comparison of the energy spectrum of the photons emitted
by an ideal radiator with the energy spectrum of the photons
emitted by a radio station. The photons from the radio station
have a spectrum that is much more sharply peaked than
the energy spectrum of an ideal radiator.

energies as a function of the photon energy (Fig. 1).*
When we form a spectrum of visible light by passing the
light through a prism, we spread the light into its
various colors (that is, into the various frequencies or
energies of the photons in the light), and thus we can see
how many photons appear at different energies (different
colors). The universal microwave background has a spectrum typical of an ideal radiator (Fig. 1). As discussed in
Chap. 8, an ideal radiator is an object that has reached
equilibrium with its surroundings, so there is no net
energy flow into or out of the object. The ideal radiato-

---

*We can also form the spectrum in terms of wavelengths or frequencies of
the photons rather than their energies.

reaches a certain temperature that characterizes the object and the amount of energy which it radiates (and absorbs) each second. Such an ideal radiator, which is in equilibrium with its surroundings, *emits* photons whose energy spectrum has a characteristic shape, like a fingerprint (see page 183).

The characteristic shape of the energy spectrum of the photons emitted by an ideal radiator allows us to disentangle and identify the universal microwave background from the other types of radiation we receive. The *shape* of an ideal radiator's energy spectrum is always the same, but the actual numbers depend on the radiator's temperature. Every ideal radiator emits photons over a wide range of energies. It emits the *maximum* number of photons at a certain energy (or frequency) that is *characteristic of the ideal radiator's temperature.* We shall call this characteristic energy E_{max}. If we double the temperature of a given ideal radiator, we shall double the energy E_{max} at which the maximum number of photons are emitted. All ideal radiators have similar spectra. This similarity in the spectral shape allows us to determine whether or not a given set of photons (light or radio waves) was produced by an ideal radiator.

By measuring the number of photons in the microwave background at many different energies, astronomers have found that we are immersed in a universal bath of photons that has an energy spectrum seemingly identical to the energy spectrum produced by an ideal radiator at a temperature of 3° above absolute zero.[1]

How do we know that this microwave radiation is a relic of the early years of the universal expansion? Let

---

[1] Unfortunately, we can not yet make measurements at energies equal to or somewhat greater than the expected peak in the spectrum of the microwave background radiation. Our inability to observe these microwaves lies in the earth's atmosphere, which contains water molecules that absorb photons with these energies. We would very much like to make these observations from outside our atmosphere to be sure that we do have a true ideal-radiator spectrum for the microwave background radiation.

us consider again the early history of the universe. Recall that the first few minutes of the primeval fireball created a huge number of photons. Until the universe reached an age of about 300,000 years, the photons tended to collide quite often with the protons, neutrons, electrons, helium nuclei, and any atoms they might form. These collisions allowed the particles with mass and the photons to exchange energy back and forth. As a result of this steady exchange of energy, the particles with mass were able to reach equilibrium with the photons among them. The photons were constantly being absorbed and reradiated in encounters with other particles, and thus the whole universe became an ideal radiator with some characteristic temperature. This characteristic temperature decreased steadily as a result of the universal expansion because the entire universe was cooling as it expanded. The high temperatures at the beginning came from the enormous densities and pressures in the primeval fireball. During the first 300,000 years of the universe the photons and the particles with mass collided with one another often enough to maintain an even distribution of energy between the photons and the particles.

Once atoms were able to form and persist, there was no longer a significant exchange of energy between the particles with mass and the photons. The photons then passed among the atoms, only rarely being absorbed or emitted by them. The universal flux of photons has therefore preserved its characteristic ideal-radiator energy spectrum from the time atoms formed to now. All the photons in the background have decreased their energy about 1,000 times since, but the energy spectrum still has the same *shape*. The particles with mass no longer are an ideal radiator; they have not been in equilibrium with the sea of photons since 300,000 A.B.E. and have undergone all sorts of clumpings and energy changes as they formed galaxies and stars. By knowing about the universal expansion,[1] we can calculate from the present average energy of the photons in the microwave background that the temperature of the universe must have been about 3000° above absolute zero in 300,000 A.B.E.* Since then, when atoms first formed in large numbers, the expansion of the universe has lowered the energies of all the photons in the microwave background 1,000 times; this has decreased the characteristic temperature of the background radiation from 3,000 to 3° above absolute zero. Three degrees above absolute zero is a very low temperature, but no matter where we turn in space, we find photons with an energy spectrum characteristic of an ideal radiator at this temperature.

SUMMARY

The early moments of the universal expansion were times of immense densities, pressures, and temperatures. During the first few seconds of the big bang or primeval fireball that began the expanding universe, all sorts of particles and antiparticles were created and destroyed in strong and weak nuclear reactions. After a few minutes, the density decreased to the point where these reactions stopped, and the basic mixture of particles in the universe had been established. Most of the particles were

---

[1] It is a remarkable fact that years before the microwave background radiation was discovered, its existence was *predicted* by a Russian-born astrophysical genius named George Gamow, and *actually* detected in 1965.

*The spectrum of the microwave background radiation now has the shape that is characteristic of an ideal radiator, with a peak at a photon energy of 5×10^{-16} erg. This peak corresponds to a temperature of 3° above absolute zero. In 300,000 A.B.E., when the background radiation stopped interacting significantly with the massive particles, the spectrum peak occurred at a photon energy of 5×10^{-13} erg, and the corresponding temperature was 3,000° above absolute zero.

then protons, electrons, neutrons, helium nuclei, neutrinos, antineutrinos, and photons. As the universe continued to expand, the photons collided quite often with other particles. These collisions tended to break apart any atoms that happened to form, and they allowed the photons and the other particles to share their energies thoroughly. However, as a result of the universal expansion, the photons' wavelengths were constantly increasing and their energies decreasing (we can think of this as the Doppler effect for a photon produced at an earlier moment in the universe's history of expansion). As a result of the energy decrease, eventually the photons did not have enough energy to break apart atoms. From then on (300,000 A.B.E. — after the beginning of the expansion), atoms formed and persisted. Millions of years later, clumps of atoms formed stars and galaxies.

The sea of photons within the universe, created in its early minutes, has remained mostly unchanged *except* for a constant increase in the photons' wavelengths since 300,000 A.B.E. These photons have *not* significantly interacted with other particles since. These photons form a *universal* microwave background radiation that was predicted to exist in 1956 and first detected with radio telescopes in 1965.

QUESTIONS

1 How did people first find out that the universe is expanding?
2 What additional evidence is there that the universe is expanding?
3 Of what is the universal microwave background made?
4 When did the universal microwave background originate?

5 How did the universal microwave background originate?

6 Could the Andromeda galaxy be made of antimatter?

7 What is the *approximate* total electric charge of our galaxy?

8 What was the big bang? When was it?

9 When did atoms first form in the universe? Why didn't they form earlier?

10 What happened to the microwave background's photons after atoms formed?

11 Did stars and galaxies form before or after the moment when the photons stopped interacting significantly with the other particles?

12
the
structure
of things

Cyril's muttering and the Dog's fight song had diverted
the crew from worrying about the dangerous Doctor Zed.
All, that is, except Julie Card, who reflected on the
information that Zenith had revealed. Puzzled by Zed's
fascination with the past, she asked, "Why is Zed so sure
that he'll find the knowledge he wants at the Permids?"

Card's question suddenly restored the heavy
atmosphere that Cyril had managed to lift briefly. Borg
decided that in situations like this it was best to share
all her information with the crew. It was do or die
together.

"Well, it's like this. Zed was always a fanatic about
oofoes. He thinks the Permids show how to get in touch
with them."

"Oofoes?" Zots asked. "Do you mean other kinds of
intelligent life?"

"Precisely, B. Z. Let me show you a tape that Zed
made some eighty years ago. Actually, it was his last
official report to the Trust."

Zenith handed Zots a small reel of hologram videotape and sat down in the captain's couch. Zots inserted
the tape in the illuminator, turned a dial, and suddenly
a clear space in the center of the cabin was filled by a
three-dimensional figure of Bokomoru Zed as he appeared

in his prime: slight but not yet stooped, intelligent rather than gnomish, with a fanatic's gleaming eye, a full head of hair, and bushy black eyebrows. Zed's image looked from side to side and intoned meaningfully:

"Oofoes: Friend or Enemy?"

The voice was so familiar to Cyril Zaki that he stopped walking around Zed's image to look at the hair (he was sure Zed wore a wig) and reclined into a vacant crash couch to listen to the mesmerizing doctor, who spoke in a commanding, faintly petulant tone that demanded attention.

"Oofoes: Friend or Enemy?" Zed repeated with furrowed brow. A long pause, and then he spoke again, slowly at first but with deepening concern.

"Although the Trust has decided that the study of oofoes is a bad risk, I submit that the well-being of humankind demands that we know all we can about nonhuman civilizations, and I refuse to be cowed and I submit this report so that in the future no one can accuse me of having failed to bring the obvious to the attention of the so-called responsible power structure."

Zaki felt his blood pressure rising. "That's Zed all right," he informed the crew. "But he's wearing a wig."

"Cool it, Cyril," said Julie Card. "Let's hear him out."

Zed continued: "The Guaranteed Trust realm includes two thousand planetary systems spread through fifty thousand cubic light years of the Milky Way galaxy. Now even our own galaxy has a hundred billion stars, and the most conservative estimate of the number of stars with planetary systems in our galaxy alone is one hundred million. Let us grant, for the sake of argument, that no one knows the exact processes by which life developed in this region of the galaxy. But this is just another way of saying that life is equally likely to spring up in any *part of the galaxy, or in any other galaxy. And if we could only find other intelligent beings, how wonderful it would be! Even though we need a certain temperature, chemical composition, and so on to produce* life, still, the very fact that we have found so many planets capable of sustaining life as we know it must constitute unarguable proof that out of a hundred million planetary systems there must be at least a million favorable to life forms. This is so obvious I can't waste my time discussing it further.

"Now, how many of the two thousand planetary systems in the Trust realm contain planets that can sustain human beings? The answer is no secret; it is about three hundred. What conclusion emerges from this analysis to all but the most constipated minds? Obviously, that we have sampled less than one-tenth of one percent of the planetary systems in our galaxy. And yet the idiots who pretend to care for our welfare do not hesitate to assure me that no other civilizations exist in the Milky Way galaxy! What effrontery! What fantastic lack of imagination!

"What are the true facts? Guaranteed Trust investigators have visited only one one-hundredth of one percent of the stars in this galaxy, and the G.T. overlords pretend to know all the answers! Not just to fool the people, either, but really to fool themselves! Each generation finds society slipping further into the morass of ignorance, and the Insuror Extraordinary does nothing. In fact, he seems to relish power without caring for its responsibilities. Humanity is headed for the junk-pile, unless we can find other civilizations that have the knowledge we need to save ourselves from our growing stupidity. They must be there! We can find them if we only try! Let us face the truth in life, fearless and unafraid, or else be content to die in the mire of ignorance. We can not afford to let the secrets of the universe pass by us. I can not stand it. I will fight, alone if I must, confronted by the upclose, constipated, so-called civilizors of the Trust and its High Assurance and its Assistant Evaluatory and the whole rotten muckery, but I will not waver. I will persevere. Now I have spoken. You must act."

Zed's image pointed a bony finger at the crew for a few seconds and then disappeared, leaving a reflective group to ponder over what they had seen and heard.

Cyril Zaki spoke first. "It's a wig, I tell you! And those eyebrows were fake, too."

As the discussion between Borg and Card, Zots and Dalby raged about him, Cyril found himself left out. Luckily, he realized that his moment of reassertion was at hand.

"Everyone into their crash couches!" Cyril yelled. "We're starting our deceleration into Thurd!"

"Blow it at me, baby," said Zots, as he started the suids moving. "We're coming down now."

A summary of what is happening in the universe should
include these three facts:

1. The universe is expanding.
2. The universe is filled with microwave radiation.
3. Matter in the universe is highly clumped into stars and galaxies.

We discussed the first two items in Chap. 9 to 11. The third point—the clumpiness of matter in the universe—is of tremendous importance; it deserves close attention, if not outright admiration. We live on solid earth and look outward at vast and nearly empty spaces strewn with an occasional star. The material within an average star like our sun has a density equal to that of water, but the average density of the material spread between the stars is 1 trillion-trillion (10^{24}) times less than this! This low density refers only to the interstellar material inside our own galaxy. Between galaxies, the density of matter is at least 100,000 times less than the density of the diffuse gas within our galaxy.

However, there is so much space in a galaxy that the interstellar gas, even at its tiny density of one atom per cubic centimeter, forms a significant fraction of the mass that is found in stars. In particular, the interstellar gas amounts to 1 or 2 percent of the total mass of our Milky Way galaxy. What makes this interstellar material important is that stars must have somehow condensed out of this type of diffuse matter. During the early stages of the universe, matter (and energy) had a fairly uniform distribution throughout space. The high temperatures and pressures in the early universe, which occurred because the universe was so much denser then, gave particles large, random velocities that tended to smooth out any fluctuations from place to place. But the universe that we see now shows tremendous *contrasts* in the degree of concentration of matter from one location to another; for example, stars are 10^{24} times denser than interstellar matter.

One of the most crucial problems in astrophysics, as yet unsolved, is to understand how these concentrations of particles ever grew out of small density fluctuations. A density fluctuation is a region in space where there are either more or less particles per cubic centimeter than in the general surrounding regions of space. The concentrations of particles formed from initial density fluctuations despite the continuous decrease in the density of matter that arose from the expansion of the universe. Although we do not know much about the concentration process, we believe that the condensations were formed in *decreasing order of size* as the universe expanded and cooled. The first important concentrations of matter must have been the clumps of gas that became clusters of galaxies. Then the individual galaxies started to condense within the cluster material. As the gas clouds shrank to galaxy size, individual stars and clusters of stars began to form within them. The condensation process that produced the stars must have left (at least occasionally) small lumps of debris; one of these is our blue-green earth. In the process of condensation through decreasing orders of size, gravitational forces played the dominant part. Strong, weak, and electromagnetic forces became as important as gravitational forces only after a certain degree of concentration had occurred; we meet them later, when we examine the results of the condensation process: the stars. Thus the sequence of the three stages of the condensation process was

Galaxy clusters → galaxies → stars in galaxies

Figure 1 represents the hierarchy of clusters in the universe; we shall discuss each of these kinds of astronomical objects.

GALAXY CLUSTERS

On photographs of distant objects, many faraway galaxies appear near one another in the sky (Fig. 2). Most of these

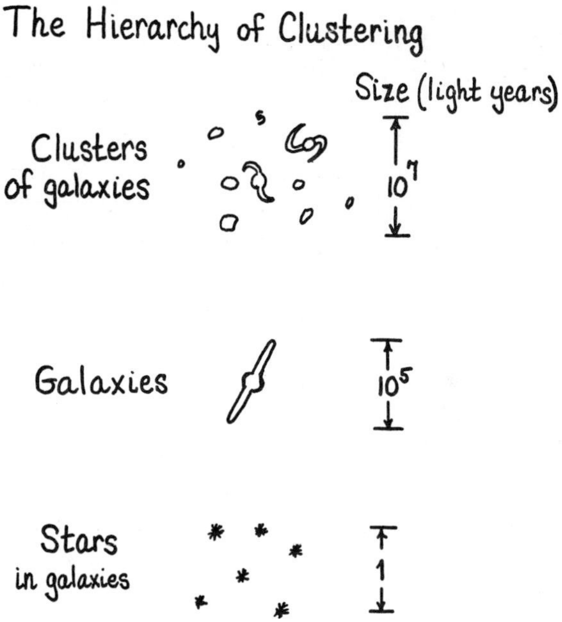

The Hierarchy of Clustering

FIGURE 1

The different sizes of clustering within the universe. The largest clusters, groups of galaxies, have diameters of 10 or 20 million light years. The galaxies in these clusters are each gigantic clumps of billions of stars, with a diameter of about 100,000 light years. Within galaxy clusters, the individual galaxies are separated by distances 10 or 20 times the size of a galaxy.

Inside a galaxy, the individual stars are about 1 or 2 light years apart, and each star has a diameter something like a million miles (less than one-millionth of a light year). This means that individual stars are separated by distances millions of times greater than the size of a star.

apparent associations of galaxies reflect a true clustering of galaxies in space. This conclusion gains support from measurements showing that the galaxies which seem to be near one another in our line of sight have almost the same velocity of recession from us. Hubble's law for the universal expansion (Chap. 9) implies that galaxies which move away from us with the same velocity lie at the same distance from us.

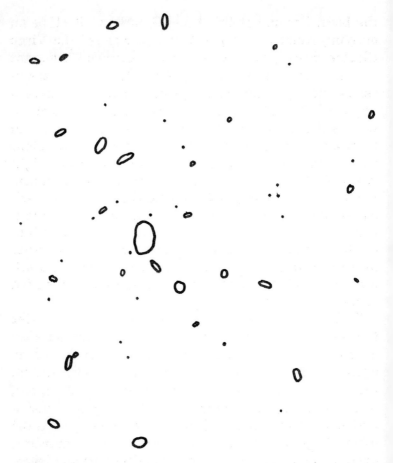

The cluster of galaxies in the constellation Coma Berenices. Each object with some extent is a galaxy, and the dots are foreground stars in our own galaxy. Each individual galaxy in the Coma cluster is about 380 million light years away, and each contains billions of stars. In contrast, the foreground stars in the picture are 100 to 1,000 light years away from us. That is, the galaxies in the Coma cluster are a million times farther away than the foreground stars.

Clusters with as many as 1,000 member galaxies have been discovered. Figure 2 shows the central parts of the famous cluster of galaxies in the constellation Coma Berenices. Our own Milky Way galaxy belongs to a small cluster with only 16 members, which we cheerfully call

the Local Group, but this Local Group may itself be an
outlying member of a great grouping called the Virgo
Cluster. Some astronomers have suggested that giant
clusters, like the Virgo and Coma clusters, are themselves members of "superclusters," but this remains
unproved. Therefore, right now galaxy clusters seem to
represent the *largest* concentrations of matter that can
be regarded as coherent blobs of material. Such giant
clusters can have diameters greater than 50 million light
years. Even the miniature Local Group, whose members
include the Milky Way, the Andromeda galaxy, and various smaller galaxies, has an extent of more than 2 million light years. The individual member galaxies within
clusters perform complicated motions around one another
because of their mutual gravitational attraction. Each
orbit for a galaxy in a cluster may require several billion
years to complete.

We believe that the formation of the clumps of matter
that became galaxy clusters must have happened after
atoms were able to form in the universe. Before then,
during the first 300,000 years of the universe's expansion,
the material in the universe was uniformly spread out,
with no density fluctuations. This homogeneous distribution occurred because photons kept colliding with the
other elementary particles that made up the universe.
Such constant collisions kept the matter in the universe
from bunching in clumps. After the density of matter in
the universe had decreased so much that the background
radiation (photons) uncoupled from the matter, then
significant clumping of matter could take place. (We
discussed this uncoupling in Chap. 10.) Once the homog-
enizing effect of the photon collisions no longer operated
on the matter, differences from place to place in the
density (mass per cubic centimeter) could persist and
grow larger. We think that such a clumping process went
on during the first billion years of the universal expansion. Regions with greater-than-average density tended
to attract more particles because these regions exerted
greater gravitational forces, and this process proceeded

faster and faster as more matter gravitated toward the denser regions. A more massive clump of matter has a stronger gravitational attraction for other matter because the gravitational force depends on the mass (see Chap. 1).

Thus we can see that once small clumps of matter (density fluctuations) has formed, more clumpiness was likely to follow because of the action of gravity. But how did the small clusters of matter, centers for future condensations, ever appear? We do not know, as we do not know many things about the universe (for instance, how it started to expand). Our own lives and the history of the world remind us that the hardest step is always the initial push; if the push comes, we can see how its effects grow and multiply. For the case of galaxy clusters, we know that during the period that began 300,000 to about 1 billion years after the big bang, clumps of matter with densities only a few percent greater than their surroundings could grow by using their gravitational forces to become giant clumps of matter—clusters of galaxies—with average densities 100 or more times greater than the material between the clumps. Nevertheless, how the first few percent of density differences needed to form the clumps ever arose remains a mystery.

GALAXIES

Within the great clouds of gas that became clusters of galaxies, the separate galaxies must have contracted from individual blobs of gas, in a way similar to the one that the clusters themselves contracted. These proto-galaxies (galaxies in the process of formation) represent a further partition of the universe into denser and less dense regions. Galaxy clusters show this partition for a distance scale of millions of light years, whereas galaxies have characteristic sizes of only 100,000 light years. To start the proto-galaxies as distinct blobs of gas, there must have been some sort of initial clumps of matter (density fluctuations) within the cluster. These fluctuations grew denser and more massive through the workings of gravitation until they became galaxies. This is the same process as the formation of galaxy clusters, but on a smaller scale of size.

The critical factor in the contraction of a proto-galaxy was probably the amount of *spin momentum* that the proto-galaxy had when it separated from the rest of the material in the cluster. We can picture the fragmentation into proto-galaxies within a cluster as something like the picture shown in Fig. 3, where the initial proto-galaxies are represented as condensations, with arrows

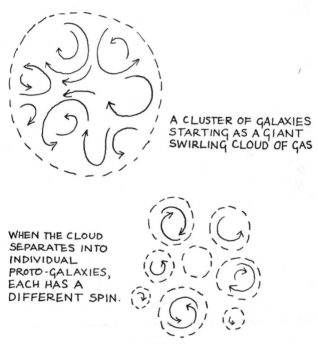

A CLUSTER OF GALAXIES
STARTING AS A GIANT
SWIRLING CLOUD OF GAS

WHEN THE CLOUD
SEPARATES INTO
INDIVIDUAL
PROTO-GALAXIES,
EACH HAS A
DIFFERENT SPIN.

FIGURE 3
A schematic representation of the fragmentation of
the giant gas clouds that became clusters of galaxies. Each blob
that fragmented out of the gas cloud could have had some
spin momentum, as we have represented by the arrows
in the figure. The total spin momentum for the entire cloud
could total zero because some of the spins of the individual
blobs (proto-galaxies) would be in opposite directions.

marking the direction of their spin. We expect that each of the proto-galaxy blobs would be rotating, as in Fig. 3. Many of the galaxies that we see now, including our own, are rotating, and they have a large amount of spin momentum. The concept of spin momentum has been found to be useful for making calculations and also reflects an intuitive feeling we have about spinning bodies. For a given object, the spin momentum is proportional to the *size of the object squared times its rate of spin*,[1] and it is important to remember that this spin momentum stays *constant* as the object expands or contracts if no forces act on it. This constancy of spin momentum is familiar to figure skaters and acrobatic divers. It implies that when a spinning body contracts, it spins more rapidly than when it was larger. Thus, for example, a high diver will contract her body to increase the rate at which she tumbles (Fig. 4). The rate of spin tells us how fast a body spins around and is usually measured in revolutions per second.

Our galaxy must have had some spin momentum when it became a separate blob of gas; this spin momentum has remained almost constant as our galaxy contracted. The constancy of our galaxy's spin momentum implies that the spin *rate* has increased because our galaxy is now much smaller than it was during its proto-galaxy stage. The rate of spin of our galaxy is about one revolution per 200 million years (this time period is sometimes called a *cosmic year*). Figure 5 shows a top view of our Milky Way galaxy and the sun's orbit around the galactic center. The

---

[1]Strictly speaking, it is the average radius of the object in directions perpendicular to the spin axis that appears in the formula for spin momentum. The amount of spin momentum S that a given body has can be written as

$$S = 2\pi \times U \times M \times R^2$$

U is the number of revolutions per second, M is the mass of the body, and R is the average radius perpendicular to the spin axis. The spin momentum of a galaxy (or any object not acted upon by external forces) remains constant. When a galaxy shrinks in radius R without losing any mass M, its rate of rotation U must increase to keep its spin momentum S constant.

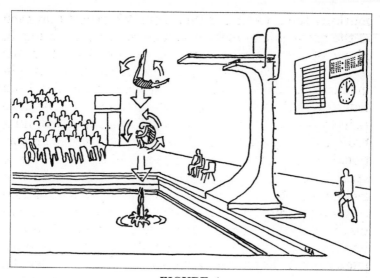

FIGURE 4
A high diver spins around as she dives. To spin faster,
she contracts her body to a smaller size.

sun's speed in its orbit around this center is about 160 km/sec. (We may compare this with the earth's speed in its orbit around the sun, which is 30 km/sec.)

The spin momentum of a proto-galaxy is important because the galaxy's spin can stop the contraction process when the proto-galaxy has contracted to a certain size. This critical size at which no further contraction will occur is reached when an average piece of the galaxy has enough kinetic energy in its rotational orbit to balance the gravitational forces from the inner regions of the galaxy. Figure 6 shows how rotation tends to expand a spinning object like a proto-galaxy. In general, particles move in straight lines at a constant speed *if no forces act on them*. Thus each piece of a proto-galaxy would move in a straight line rather than in a circular orbit, *if* there were no gravitational forces to keep the pieces in their orbits. This tendency of a particle in orbit to move in a straight line, rather than circling in orbit, is sometimes

258

FIGURE 5
A schematic drawing of our Milky Way galaxy as seen from above,
showing the spiral arms of the galaxy and the sun located about
30,000 light years from the galactic center. The sun orbits around
the center at a speed of about 160 km/sec. Since the orbit has
a circumference of about 100,000 light years or 10^{18} km,
the time for the sun to go around the galactic center once
is $6\frac{1}{4} \times 10^{15}$ sec or about 200 million years.

called *centrifugal force*, though it is *not* a real force; it
simply reflects the fact that particles move in straight
lines unless forces act on them. Within a spinning proto-
galaxy, the central regions hold together the pieces of
gas because they contain a great deal of mass that exerts
a large gravitational force. Eventually, each blob of gas
in the proto-galaxy reaches a distance from the center
and a rate of spin around the center that reflects a delicate balance between the centrifugal forces and the forces
of gravitation.

The point at which the galaxy's spin will stop the contraction process depends on the amount of spin momentum in the original proto-galaxy. The spin momentum,

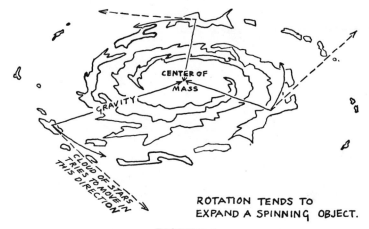

ROTATION TENDS TO
EXPAND A SPINNING OBJECT.

FIGURE 6

If a body rotates, it tends to expand because each particle
in the body tends to move in a straight line unless some force
that keeps the particle in a curved orbit is applied. This
tendency for particles to move in a straight line is sometimes called
centrifugal force. Some additional force must counteract the
centrifugal force to prevent a rotating body from expanding.

which varies as the size of the object squared times the
rate of spin, stays the same as the proto-galaxy shrinks
and spins faster. Eventually the spin rate becomes large
enough to give each point of the galaxy just enough
speed to balance the gravitational forces, and the contraction stops.

Spiral galaxies seem to have acquired a large amount
of spin momentum when the became proto-galaxies and
separated from the rest of the galaxy cluster. These
galaxies are highly flattened along the spin axis (Fig. 7)
because along this axis there is no spin momentum to
support the galaxy's material against the gravitational
pull from the central regions of the galaxy. However, the
galaxy's rotation in the plane that is perpendicular to
the axis of spin enables the galaxy to remain spread out
in these two directions that are not along the spin axis.

In spiral galaxies, the rotation of the whole assemblage
of stars helps to support the galaxy against the inward
gravitational pull from the massive central regions. As

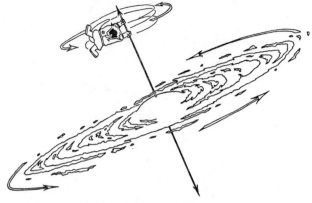

FIGURE 7
A spiral galaxy is highly flattened along the spin axis (axis
of rotation). The effect of the galaxy's rotation is to support
the galaxy against gravitational forces in the directions
perpendicular to the spin axis, but this effect does not operate
along the axis of rotation. The result, when the galaxy has
reached an equilibrium balance between the rotation effects
and the gravitational forces, is that the spiral galaxy is
much larger in directions perpendicular to the spin axis
than it is along the spin axis.

part of this rotation, each star in a spiral galaxy travels
in a nearly circular orbit around the galactic center. We
may consider the entire galaxy as a rotating blob, or we
may examine each star as a separate particle rotating
around a common center. In either view, the spin momentum that arises from the stars' motion keeps the outward
centrifugal forces and the inward gravitational forces in
balance.

For a spiral galaxy, rotation provides support in the
two dimensions perpendicular to the axis of spin. What
keeps the galaxy from flattening itself to an incredible
thinness in the direction along the spin axis? The support in this direction arises from the individual random
motions of the stars parallel to the spin axis. The gravitational field of the galaxy tends to pull a star toward the
galaxy's central plane, where the density of stars is
greater than in regions away from the plane. However,

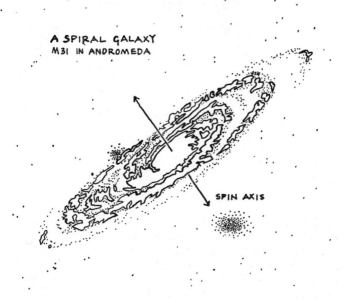

A SPIRAL GALAXY
M31 IN ANDROMEDA

SPIN AXIS

FIGURE 8

stars in a galaxy are so widely spaced that collisions
among them occur only very rarely. Thus a star can pass
right through the median plane and head out the other
side because of its own momentum. This oscillating
motion in the direction perpendicular to the central
plane of a galaxy is something like the highly elliptical
orbit of Halley's comet around the sun. Each time the
comet returns to our vicinity it is headed almost straight
at the sun, increasing its velocity steadily. But instead
of running into the sun, the comet swings around it and
decelerates as it returns to the far reaches of its orbit. In
a spiral galaxy, stars are attracted toward the plane of
the galaxy (which replaces the sun in the model based on
Halley's comet); as they orbit the galactic center, they
pass right through the plane, decelerate, and eventually
reverse direction to return once more, as shown in Fig. 9.

Thus, within a spiral galaxy the motion of each star
has two chief components. First, all the stars orbit the
center of the galaxy, in nearly the same plane and in the

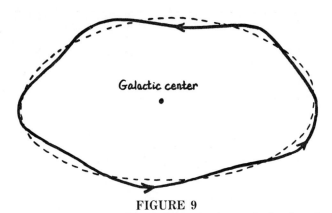

FIGURE 9

The gravitational forces from other stars in a spiral galaxy tend
to pull an individual star toward the central plane of the galaxy.
Such forces cause the star to oscillate through the plane as it
orbits the galactic center. The smaller orbit is superimposed on
what would be a circular orbit (dashed line) to produce a
total motion like that of a horse on a merry-go-round.

same direction, traveling at speeds of several hundred
kilometers per second. This rotational velocity decreases
from the inner to the outer parts of the galaxy, but the
overall galactic rotation provides the support needed to
keep the galaxy more than 100,000 light years in diameter.
Secondly, superimposed on these circular orbits around
the center we find motions perpendicular to the plane of
the galaxy. These oscillations up and down through the
plane are smaller in size and velocity than the circular
orbits of stars around the galactic center. Each star per-
forms a motion analogous to that of a horse on a merry-
go-round (Fig. 9). An average star may reach a height of
600 light years above or below the central plane and will
pass through the plane at a speed of 10 or 20 km/sec.
Right now our sun is located within 100 light years of the
central plane of the Milky Way spiral galaxy.

Another class of galaxies, about as numerous as spiral
galaxies, are called "elliptical" (Fig. 10). Elliptical galaxies are more blob shaped and far less flattened than
spirals. We may think of ellipticals as galaxies in which
the overall rotation of the proto-galaxy did not over-

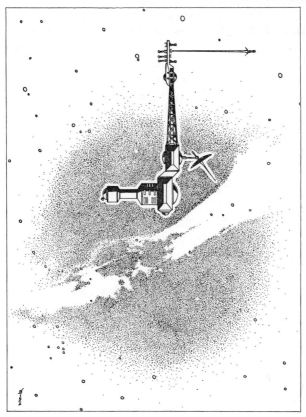

FIGURE 10
An elliptical galaxy does not show the large degree
of flattening that a spiral galaxy does.

shadow the random velocities of the component stars.
Elliptical galaxies do rotate, but if we once again consider each star's motion as an orbit around the center
plus a random motion, we find that the individual random velocity is about equal to the rotational velocity. In
spirals, the star's velocity from the general rotation far
exceeds the star's random velocity, and this excess gives
spiral galaxies their characteristic flattened form.

In spiral galaxies, a few percent of the galaxy's mass
still float among the stars in the form of gas atoms or
dust grains. We think that each dust particle consists of

a few million atoms, mostly carbon (which bind quite firmly into graphite), layered with ice and other impurities on their surfaces. Elliptical galaxies, on the other hand, contain almost no gas and dust. We may conclude that star formation in ellipticals has proceeded more efficiently than in spirals, leaving no residue of gas and dust to be observed at the present time in elliptical galaxies.

In both classes of galaxies, elliptical and spiral, which, taken together, comprise most of the galaxies that have been observed,[1] the rotation of the galaxy plus the random motions of the individual stars stopped the process of gravitational contraction within the proto-galaxy at a point where the galaxy still had an immense size, like that of the Milky Way galaxy. As the contraction process stopped, individual stars began to condense out of the gas that formed the proto-galaxy. This was the third stage of the condensation process:

$$\text{Galaxy clusters} \rightarrow \text{galaxies} \rightarrow \text{stars in galaxies}$$

STARS

We have seen that huge clumps of gas formed clusters of galaxies, which in turn fragmented into individual proto-galaxies. As the proto-galaxies contracted, they in turn fragmented into proto-stars, clumps of gas that contracted further under the influence of their own self-gravity. At each stage of the fragmentation process that we described, the average gas density within a clump—proto-cluster, proto-galaxy, or proto-star—was increasing steadily: Each step in the condensation process represents a sharper division of space into (1) vast, almost empty regions and (2) much smaller regions of relatively dense material.

---

[1] The remaining galaxies are called "irregulars."

All the stars observed until now have a relatively small spread in their *masses*. The most massive star that has been detected in our galaxy has 70 times the sun's mass; the least massive has about one-tenth the sun's mass. The total range in stellar masses from largest to smallest is therefore less than 1,000 times. (By contrast, stars vary by a factor of 10^8 or more in their true brightnesses.) The relatively small variation among the masses of all the stars must indicate that clumps of gas of a certain range of mass had a better chance of condensing into proto-stars as the galaxy contracted. In fact, astrophysicists have made rather detailed calculations of how proto-stars contract, and these calculations show that within a galaxy, clumps of gas in the range from 100 to one-tenth solar masses are indeed more likely to contract than smaller or larger clumps.

While the contraction of proto-stars down to star size was well underway, some of the proto-stars must have left behind some gas and dust that formed the subcondensations of a planetary system. Our own solar system nestles close to our parent, the sun: The distance from the sun to the earth is 200,000 times less than the distance from the sun to the next nearest star. The distance from the sun to Jupiter, the most massive planet, is just 5 times greater than the distance from the sun to the earth, and the distance from the sun to Pluto, the farthest planet, is about 40 times the distance from the sun to the earth. If the sun were the size of a grapefruit, the earth could be a speck of dust 50 ft from it and Jupiter a ladybug orbiting the grapefruit at a distance of 250 ft (Fig. 11). On this scale, the nearest star to the sun would be another grapefruit 2,500 miles away, and the Milky Way galaxy of stars would have a diameter of 90 million miles, which is about the true distance from the earth to the sun.

We may conclude from this model that the proto-sun must have left behind the gas and dust that formed into planets only during the final stages of its condensation as a proto-star. The same conclusion applies to the nearby

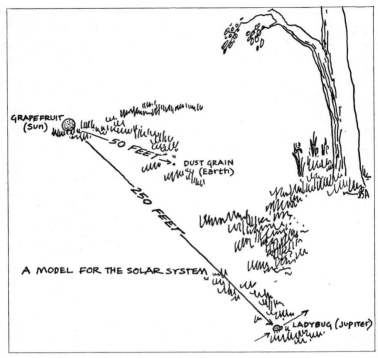

FIGURE 11

We can make a model for the solar system by using a grapefruit
to represent the sun, a dust grain for the earth, and a ladybug
for Jupiter, the largest planet. All the planets orbit around
the sun in the same direction and in almost the same plane,
so the entire solar system is relatively flat. The planet nearest
the sun, Mercury, would be about 20 ft from the grapefruit
in this model, and the farthest planet, Pluto, would be about
2,000 ft from the grapefruit. The nearest star to the sun
would be 2,500 miles away.

star system called Barnard's star, where astronomers
have detected a body with a mass like that of the planet
Jupiter (300 earth masses) in orbit close to a star somewhat cooler than our sun but much like it. Astronomers
believe that many stars have planets in orbit around
them and that these planetary systems are the not-so-
unusual residue of the proto-stellar material, left behind
as the stars contracted to their present size.

We mentioned earlier that the contraction of proto-galaxies seems to have been stopped by the spinning of the proto-galaxies. Although most stars are rotating, the amount of their spin momentum is not large enough to have stopped them at their present sizes during the contraction process.[1] The condensation of proto-stars stopped not because of rotation, but because the stars began to turn on by liberating energy in nuclear fusion reactions. As we discuss in the next chapter, this liberated energy presses outward to oppose the inward forces of self-gravitation and it is this outward pressure that keeps stars like our sun going and glowing.

★　★　★　★　★　★　★　★　★　★

Let us consider what might occur if a proto-galaxy were formed with very little spin momentum but with the same size and amount of material (mass) as an ordinary proto-galaxy. Such a low-spin-momentum proto-galaxy could shrink to a small size (astronomically speaking) before the condensation process was stopped by an increase in the rate of spin. A proto-galaxy with a spin momentum 10 times less than that of our own galaxy could contract to a size 100 times smaller than the Milky Way before a balance was reached between the gravitational forces and the motions of the galaxy's pieces in orbit. The contracting galaxy would then have an average density of material thousands of times greater than the average density of matter in our own galaxy. Such a galaxy would be quite different from what we consider and ordinary galaxy.

Two American astronomers, George Field and Stirling Colgate, have suggested that such highly condensed

---

[1] The sun, for instance, rotates about once every month.

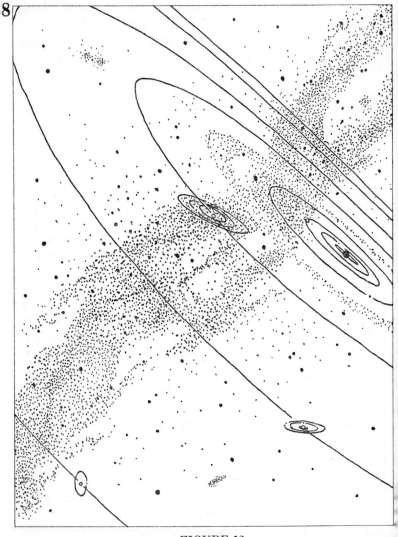

FIGURE 12
The solar system.

proto-galaxies are the sources of the remarkable *quasi-stellar radio sources* that were first discovered in 1964. These objects, which are now often called *quasars*, derive the quasistellar part of their name to the fact that on

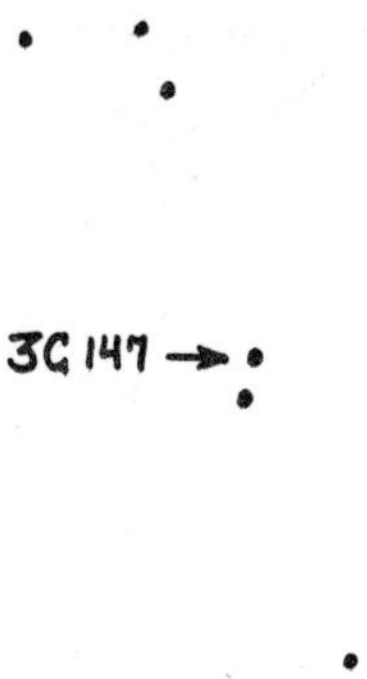

FIGURE 13

A picture of the quasar called 3C 147 shows the quasar (indicated by the arrow) looking remarkably like a star. However, the stars in the picture are 100 to 1,000 light years away, and the quasar is apparently more than *4 billion* light years away. The quasar's red shift indicates that it is moving away from us at a speed of 75,000 miles/sec, almost half the speed of light.

photographs they look like points of light, similar to stars (Fig. 13). Galaxies, in contrast, are large enough to have a fuzzy appearance even when they lie at enormous distances from us. Quasars also emit large fluxes of radio waves and thus are called quasistellar *radio* sources, even though they are also shining in visible light. By studying the Doppler shifts of the light from quasars and then applying Hubble's law for the expansion of the universe, astronomers have found that most quasars are at distances of billions of light years! The most distant object now known is a quasar at a distance of about 10 billion light years. This mind-bending separation between the quasar and ourselves means that the light we observe now must have been produced in the quasar at a time when the solar system had been formed. The speed of the quasar's recession from us is so great (in many cases, more than half the speed of light) that photons which were emitted as ultraviolet light now appear to us as red![1] Finally, the quasars must have generated fantastic

---

[1] See page 158 for the correct formula for this Doppler effect.

amounts of energy because they are not particularly faint objects (although none are bright enough to be seen without a telescope), and yet the quasars lie at astounding distances from us. We can calculate that most quasars are 10,000 or 100,000 thousand times more brilliant than the ordinary galaxies with the greatest true brightnesses.

To explain how quasars reach such a huge true brightness, Field and Colgate considered the fact that a proto-galaxy which contracts to a small size (such as 1,000 light years across) would be not at all like the Milky Way galaxy, which has a diameter of 100,000 light years. The large density of material within the highly contracted galaxy would produce a high rate of collisions between the clumps of gas that were contracting into stars. These gas clumps would tend to stick together after colliding with one another. As a result, the stars in a contracted galaxy could be much more massive, on the average, than those in an ordinary galaxy. Instead of the average star having about half a solar mass (as in our own galaxy), we might find the average star in a dense proto-galaxy with 10 or 50 solar masses.

Within a star, energy is liberated at a rate that varies approximately as the *cube* of the star's mass. More massive stars generate much more energy per second and thus are much brighter than less massive stars. The total energy that a star can liberate from its nuclear "fuel" varies in direct proportion to the star's mass, and the star's energy-liberating lifetime varies approximately as the *inverse* square of the star's mass. This means that massive stars are much brighter than less massive stars (by the cube of the mass ratio), but they have much shorter lifetimes. A galaxy that contains a great number of massive stars could be extremely luminous for some millions of years, although it would exhaust its energy sources and sink into dimness much faster than an ordinary galaxy whose stars emit light for billions of years. This may be the way that quasars achieve such high rates of energy liberation, but much more intensive work is needed to verify or disprove the theories of Colgate and

Field. Two British astronomers, Fred Hoyle and Geoffrey Burbidge, have taken the suggestion of clumping one step farther: they have proposed that a quasar consists of *one* supermassive star with 100 million solar masses![1] The British astronomer Donald Lynden-Bell has suggested that at the center of each quasar is a gigantic *black hole* with 100 million solar masses and that matter falling toward the black hole produces the quasars' radiation in the last moments before the matter disappears forever. (Suggestions have been made that *every* galaxy has a black hole at its center, which was the seed around which the gravitational clumping and contraction started.)

The amount of energy that quasars apparently emit as light and radio waves is so mind boggling that something may be fishy with our conclusions about these sources. The American astrophysicist Halton Arp has pointed out that many quasars are located on the sky in places surprisingly close to ordinary, much closer galaxies. Arp believes that quasars are in fact physically associated with these galaxies and close to them in space. The galaxies are "only" 50 to 500 million light years away, rather than the billions of light years' distance that we assign to quasars on the basis of Hubble's law for their large Doppler shifts. If quasars are indeed 10 or 100 times closer to us than Hubble's law would imply from their enormous velocities, then they would have only one-hundredth to one ten-thousandth of the true brightness we have assumed so far. This is only as bright as a very bright galaxy. Recall that according to the brightness law (page 205), the true brightness of an object which we observe to have a given apparent brightness will vary inversely with the square of the object's distance from us.

There is big trouble, though, in accepting the conclusion Arp suggests: that quasars are much closer to us than their immense velocities would indicate. Namely,

---

[1] The relationship that links a star's rate of energy liberation to the *cube* of the star's mass does not apply to stars with masses greater than about 50 times the sun's mass.

how can we explain these fantastic velocities, half the speed of light or greater, that we measure from the Doppler shifts of the quasars' photons? The big bang that started all the universe expanding has taken 15 billion years to make one huge clump of matter recede from another clump with half the speed of light, where the clumps are separated by 10 billion light years. No energy-producing mechanism that astronomers consider likely can accelerate relatively nearby large clumps of matter to anything like half the speed of light. If there were an energy source capable of accelerating a quasar to these speeds, it would probably blow the quasar apart—like trying to accelerate a racing car with a hydrogen bomb. Thus quasars present a choice of difficulties: Either we can try to explain how quasars can generate energies thousands of times greater than those of ordinary galaxies, or we can assume that quasars are actually not so far away and then try to explain how they could move so fast, and always away from us. Modern astronomy has produced several problems that, like quasars, challenge our accepted notion of how the universe works. Quite possibly, objects like quasars will eventually lead people to find new laws for the physical behavior of matter in the universe because conditions within them are so immensely different from those we can duplicate in laboratories that they may involve whole new processes as yet unknown.

SUMMARY

Matter in the universe is clumped into clusters of galaxies that are made of hundreds of individual galaxies which in turn are made of billions of individual stars. These clumps of matter could not form from the homogeneous, diffusely spread-out particles in the early universe while photons were still interacting strongly with the particles. But after 300,000 years of the universal expansion, the formation of atoms ended this interaction, and clumps could start to form and grow. During the first billion years following the formation of atoms, gravitational forces tended to magnify density contrasts to produce separate blobs of gas. Regions that happened to be a little denser than the average attracted more and more particles gravitationally, and this process continued until large variations in density from place to place occurred. These denser clumps became clusters of galaxies, and within the proto-clusters individual proto-galaxies formed a further clumping. Some of the clumps became flattened spiral galaxies; others became more spherical elliptical galaxies. In both kinds of galaxies a further subclumping formed proto-stars that later condensed into stars, occasionally leaving behind planetary systems.

Quasistellar radio sources, or quasars, look pointlike (as do stars), but they have tremendous red (Doppler) shifts that imply huge velocities of recession and hence huge distances from us. Because quasars do not appear especially faint, their true brightnesses, both in light and radio waves, must be fantastic, if they really are at distances of billions of light years. If we adopt the alternative, that quasars are not at the immense distances indicated by their red shifts, we must admit that no explanation of their enormous velocities seems feasible.

QUESTIONS

1 What are the largest agglomerations of matter in the universe?
2 Of what are they made?
3 What kind of forces produced these agglomerations? What kinds of forces are holding them together now?
4 What happened to these agglomerations as they contracted?

5 What is a proto-galaxy?

6 If an object has a certain spin momentum and is spinning at four revolutions per second, how fast will it spin if it contracts to one-half its original size?

7 When proto-galaxies fragmented out of the original agglomerations, were there already stars in them?

8 What kinds of forces tend to make proto-galaxies contract to smaller and smaller sizes? What resists this contraction?

9 What kinds of forces cause proto-*stars* to contract? What resists this contraction?

10 What is unusual about quasars? What is usual about these objects?

13
ashes
to ashes

Preprogrammed thrusts of power brought the Top Dog
smoothly into a parking orbit around Thurd, and the
crew found themselves once again lucid and mobile,
pleased that one more psychorama had ended without
apparent damage to their fragile psyches. Zots and Dalby
began to speculate on the nature of the planet beneath
them, but Cyril Zaki went to the center of information.

"What kind of a planet is this, Zenith?" he asked.

"Well, the computer rates it Baaaad."

This judged Thurd rather a doubtful risk. The "B"
meant that its atmosphere was sufficiently respirable, and
the four "a's" indicated a reliable water supply, acceptable
temperature range, no deadly forms of bacteria, and
seismic activity below the danger level. But the "d"
meant that the planet's star was a highly variable source
of energy, capable of flaring up at any moment and
zapping the unwary visitor with a flood of x-rays and
ultraviolet radiation. Cyril didn't like visiting stellar
systems with a "d" at their code end. "Give me a bunch
of "c's" every time," he liked to say. "That one "d" can
kill you for sure. Why risk so much for so little?"

"I must point out, though," said Borg, "that our
information is considerably out of date."

"How long since the file was inputted?" asked Julie Card.

"It's not quite clear, but our last complete update came in five hundred orbits ago."

"Five hundred! And not much known then!" Zaki commented. "Zenith, this is too big a risk for you to be investigating."

"No, Cyril, there are times when even the greatest risks must be spun off the tape disk. I don't see any other way to get Zed to reveal himself; it must be me that he wants to challenge. How do we get down to the Permids?"

"Well, let's see now. If we knew where the Permids are, we could send a team in the darter."

"Good thinking, Captain. It happens that our Thurd file does give a fair description of what to look for: three humps of rock beside a mighty river."

"Good risk, your E, we can handle that right away."

This was the sort of project at which Captain Zaki excelled. He banked the Dog into a pole-to-pole orbit and let the planet revolve beneath him, scanning its surface with a high-powered optical system left from the days of exploration. Zaki found the task so congenial that he began to hum a snatch of "The Good, the Bad, and the Uninsured," an almost forgotten tune. Several hours passed in this fashion.

Don't call me a good risk, don't call me doubtful,
Just give me insurance and I'll give you a mouthful
. . . .

"Triple Premium! There they are!"

Zenith Borg peered through the viewer. "Good risk, Cyril, you've made it once more. Let's get going—how many can ride in the darter? Three? All right, you and who else? Who's the best at hand-to-hand? Julie? All right, let's not waste time."

Leaving the ship's controls in the capable hands of Bernie Zots, Cyril lowered himself into the underbelly of

the Dog alongside Card and Borg, flipped down the
hatch, and disengaged the darter from its mothership
for a glide path to the surface of Thurd. As they
approached the edge of the atmosphere, the starkly barren
nature of the planet became strikingly apparent.

"Mother of G.T.! Not much to smoke down there, eh?"

"No, Cyril, but there must have been some sort of
ecosystem here once. Some kind of people made those
Permids."

By this time the darter was close enough to the
Permids for its occupants to see them clearly: tetrahedral
forms, rock piled on rock in systematic style, reaching
a hundred times a woman's height into the sky.

"Bring it down in the sand there, can you, Cyril?"

"Good risk."

The darter bounced once, rolled a while, and came to
an easy stop near the riverbank.

"Get your blaster on, Cyril, and you, Julie, see what
you can do with this."

Zenith presented Julie Card with a cylinder longer
than her arm and twice as thick.

"It's a sort of optical blaster left over from the
Weald Wars. You point it this way, press that blue stud,
and it's supposed to vaporize everything within fifty
meters in that direction. Don't use it unless you have to."

"Right ahead," Julie said, and the three Investigators
made their way to the nearest Permid, where the solid
rocks revealed a passageway into the black hole of the
structure's insides.

Every morning the sun rises and starts another day of radiant energy. Without the sun's visible light and its warming infrared light there would be no life on earth or nearby. In fact, even a change of a few percent in the amount of energy produced each minute by the sun would drastically affect our lives on earth. How does the sun manage to produce energy at nearly the same rate, day after day, for millions upon millions of years? The answer lies in the power of nuclear fusion, which we discussed in Chap. 3.

The nuclear fusion processes that occur in the sun are a combination of strong, weak, and electromagnetic interactions. These processes involve collisions that combine two individual nuclei to form another kind of nucleus. During such a process, some of the energy of mass of the original nuclei changes into energy of motion. This energy of motion is carried away from the original collision by leptons[1] and photons that are produced in the fusion process and by the additional kinetic energy of the resultant nuclei. The particles involved in these processes collide with other particles and spread the liberated energy of motion through the sun until it reaches the surface of the sun. At the sun's surface some protons and electrons but a *huge* number of photons are ejected. The photons reach us in the form of light and heat; the charged particles (protons and electrons) ejected from the sun form the "solar wind."

We can regard protons as the nuclei of hydrogen atoms, so the word "nuclei" includes protons, helium nuclei, and the nuclei of all the elements heavier than hydrogen or helium. (The sun and most other stars consist of 99 percent hydrogen and helium nuclei by mass.) Inside stars, temperatures are so high that all the atoms have their electrons completely knocked loose from the nuclei as a

---

[1]These leptons are electrons, positrons, neutrinos, and antineutrinos, which we discussed in Chap. 5.

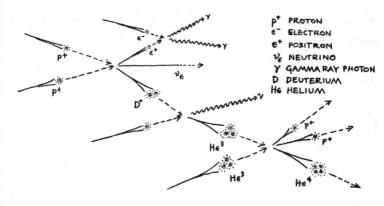

THE PROTON-PROTON CYCLE

FIGURE 1

result of the particles' collisions with each other.[1] Thus, although there are as many electrons (negative charges) as there are positive charges in the sun, the electrons are not in orbits around the nuclei. Since electrons are not subject to strong forces (see Chap. 4) they do not interact as readily as hadrons, so we must look first at the interactions of hydrogen and helium nuclei to find the sources of the sun's energy.

The fundamental series of energy-liberating processes in the sun and in most other stars is the *proton-proton cycle,* which was shown in Fig. 9, Chap. 3 (page 76). Figure 1 summarizes the three steps of the proton-proton cycle. Taken together, the nuclear reactions in these three steps have the effect of fusing four protons into one He^4 nucleus, and this fusion transforms about 1 percent of the four protons' energy of mass into 4×10^{-5} erg of energy of motion. This energy of motion is carried off mainly by the two positrons, two neutrinos, and two photons that are produced during the proton-proton cycle. After this energy is transmitted to the rest of the particles

---

[1]The energy needed to knock the electrons loose is much less than the energy needed to make two nuclei fuse when they collide.

in the sun by collisions with them, some of the energy finally reaches us here on earth because the high temperature of the sun's surface layers causes photons to be produced and radiated away. The 4×10^{-5} erg of energy of motion that is liberated each time the proton-proton cycle occurs is less than one ten-thousandth of the energy a fly uses to take off. However, since about 10^{38} proton-proton cycles occur in the sun every second, the total amount of energy of motion that is released in this way is huge: 4×10^{33} ergs/sec. (For comparison, people on earth consume about 4×10^{18} ergs/sec as they carry on with civilization.)

Protons (hydrogen nuclei) are the most abundant kind of nuclei in the universe; helium is the second simplest and second most abundant element. It is hardly surprising that stars liberate most of their energy by turning hydrogen nuclei into helium nuclei since they are using the most available nuclear fuels. Most stars have temperatures of about 13,000,000° above absolute zero [20,000,000° Fahrenheit (F)] in their interiors, where they are releasing energy through the proton-proton cycle. However, the surface layers of the stars are much cooler, with temperatures ranging from 2000 to 80,000° absolute (about 3200 to 145,000° F). Temperature in a gas measures the average energy of motion per particle. If two gaseous objects have the same temperature, their constituent particles have the same average energy of motion. At temperatures like those in the sun's center, protons have speeds large enough to overcome their mutual electromagnetic repulsion and can fuse to make helium nuclei. The ability of protons to fuse depends on the temperature in a striking way. Nuclei with larger electric charges have a stronger mutual electromagnetic repulsion, so it is much harder to fuse helium nuclei to make heavier nuclei like carbon than it is to fuse hydrogen nuclei (protons) into helium nuclei. The fusion of helium nuclei requires higher temperatures (more energy of motion per particle) than the fusion of hydrogen nuclei through the proton-proton cycle.

The centers of stars like the sun function as their own thermostats: not only do they liberate vast quantities of energy, but they are also self-regulating energy generators. The high temperature within the central region of a star arises from the great pressure of the material being drawn together by self-gravitation. The pressure makes the particles move faster, and the faster they move, the more likely they are to undergo nuclear fusion. The highest temperatures occur at the star's center, where the pressure is greatest. But the energy that is liberated in the fusion processes pushes outward from the center as the newly liberated energy of motion given to the particles there is shared with the other particles through millions of collisions. This liberated energy of motion creates a counterpressure that opposes the tendency of the star to contract from the forces of self-gravitation. The sun and other stars have reached a careful balance between the inward pressure of their gravitational forces and the outward pressure of the energy of motion liberated by the strong, weak, and electromagnetic nuclear reactions in their centers. The balance of pressures makes the inside of a star self-adjusting. If the inward forces should grow a little stronger, they will pull the central regions together slightly, which will *increase* the rate of nuclear reactions because the greater pressure on the interior will increase the temperature. The increased temperature will increase the rate of nuclear fusion reactions, and the increased energy of motion liberated by these reactions will push outward and cause the star's interior to expand slightly. This will reduce the pressure and temperature, so the rate of nuclear reactions will return to its former value, just sufficient to balance the inward gravitational pressure.

Conversely, if a little extra energy should be liberated, it will push outward and make the central regions slightly less dense. This will *decrease* the pressure and temperature, and the decrease in temperature will slow down the rate of energy liberation. This in turn allows the inward gravitational forces to restore the original balance of pressures.

The energy liberated deep in the sun's interior seeps outward by means of many collisions. Because of the high density in the sun's central regions (about 100 times the density of water), the energy of motion liberated in the proton-proton cycle can not escape at once but must be diffused to the other particles that form the sun. Similarly, the photons that appear in step 2 of the proton-proton cycle (Fig. 1) can not escape directly. Instead, they constantly change their energy as they collide millions of times with the nuclei and electrons in the sun's interior. (The neutrinos *do* escape directly from the sun's insides, but because almost all of them pass right through the earth too they are not a form of energy that is useful to us. The neutrinos carry off only a small fraction of the energy of motion liberated in the proton-proton cycle.) Photons produced by nuclear interactions in the sun's interior take about 1 million years to reach the sun's surface. From there, they stream out into space in straight lines, carrying away most of the energy produced in the sun's interior.[1]

The self-governing energy cycle going on in the centers of stars like the sun can continue to fuse hydrogen nuclei into helium nuclei for billions of years. From geologic records, and from what we know about the theory of nuclear fusion, the sun has been producing energy at approximately its present rate for the last 4 or 5 billion years. Because evolutionary processes occur over many generations, one requirement for life to develop on earth or some other planet is a relatively constant source of light and heat so that the gains made by natural selection will not be wiped out by fluctuations in the light and heat

---

[1]Because a photon making its way to the sun's surface constantly changes its energy, it is not quite accurate to refer to it as one photon; instead, what really happens is that the photon is constantly destroyed, and another appears with a somewhat different energy.

at the planet's surface. Most stars seem to provide such a
constant source, at least for the last 100 years or so that
we have studied them in detail. However, an interesting
wrinkle in this story may be the additional push to evolve
that would be exerted on living organisms if the sun
changed its true brightness. Some astronomers have suggested that the sun's "usual" true brightness is about
50 percent larger than its present value and that a
decline in luminosity occurred several million years ago,
ushering in the last great series of ice ages. Could it be
that increasing coldness on earth caused primitive men
to emerge as an effective species at about this time?
Like many scientific problems, this question remains
undecided.

When the sun first turned on and began nuclear fusion
reactions, it contained some helium, about one He^4
nucleus for every 16 hydrogen nuclei (protons). (This
helium was made from protons by nuclear reactions
during the big bang.) After 5 billion years of fusion
reactions, the sun has increased the percentage of helium
nuclei to the point where there is now 1 helium nucleus
for every 10 hydrogen nuclei in the sun. The sun can
continue to produce energy of motion by fusing hydrogen
into helium (the proton-proton cycle) for another 7 billion
years, more or less. As hydrogen gradually becomes
scarcer, the sun will contract. This contraction will make
the sun's inner regions a little denser and hotter, and
this will produce a faster *rate* of nuclear fusion to compensate for the reduced fraction of hydrogen. Eventually
(in just a few billion years), almost all the hydrogen in
the sun's central regions will have been fused into
helium. As the central regions of the sun grow smaller
to produce higher temperatures, the outer layers will
expand and grow cooler because the last burst of liberated energy that occurs as the star evolves into its new
phase pushes the star's outer envelope away from the
core in all directions. After a few thousand years, the
star reaches a new state of equilibrium, the red giant

phase. During this phase, which can last for hundreds of millions of years, the sun will have a dense core of almost pure helium, and it will liberate energy in a shell just outside this helium core (Fig. 2). The shell will contain some hydrogen nuclei and will have a temperature that is high enough (about 100,000,000° absolute) to liberate energy quickly from the ever-diminishing store of hydrogen nuclei. The sun's enlarged outer layers will be cooler (about 3000° absolute) than they are now (about 6000° absolute), so the sun will appear to be red rather than yellow.

All the stars that we see shining produce their energy by fusing four protons into an He^4 nucleus, over and over, trillions of trillions of trillions of times each second. In most stars this fusion process occurs through the proton-proton cycle, but some stars use a more complicated chain of strong and weak nuclear reactions to obtain the same result: energy of mass is liberated as energy of motion.

We classify stars by two main characteristics: their surface temperatures and their true brightnesses. Figure 3 shows this kind of classification, which reveals that most stars fall not just anywhere in this diagram but on a

The Sun As a Red Giant Star

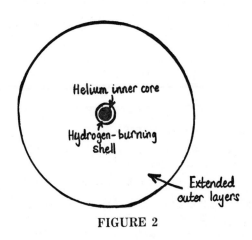

FIGURE 2

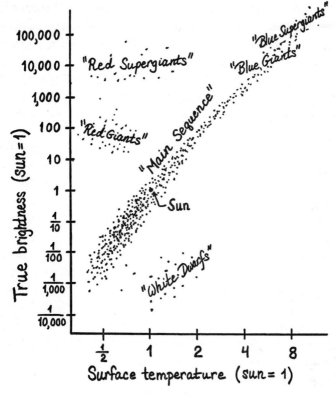

FIGURE 3
We classify stars by their surface temperatures and their true
brightnesses; the figure shows this classification. Most stars,
including the sun, have surface temperatures and true brightnesses
that place them on the main sequence region of this chart,
rather than all over the map. Other stars appear in the
various regions designated by the names of the types of stars,
such as red giants or white dwarfs.

limited region called the "main sequence." The sun, an
average sort of star, also belongs to the main sequence.
Stars that fall in various regions of this temperature-
true brightness diagram have exciting names like "blue
giants," "red giants," "white dwarfs," "red supergiants,"
and so on. But they all liberate their energy by fusing
protons into He^4 nuclei.

Blue giants are large, hot stars that burn themselves out relatively quickly. They have short lifetimes because the rate of energy liberation in a star varies approximately as the *cube* of the star's mass, and the available supply of nuclear fuel (protons) varies in direct proportion to the star's mass. Thus a star's lifetime is approximately proportional to 1 over the square of its mass. Blue giant stars are those with the largest masses. An extreme blue giant (called a blue supergiant), such as the bright star Rigel in Orion's foot, has a mass about 20 times the sun's mass. Rigel has a surface temperature of about 18,000° absolute (the sun's surface temperature is about 6000° absolute) and a true brightness about 10,000 times the sun's.

Red giants and red supergiants have enormous sizes, but their surface layers are cooler and much more extended than the sun's or a blue giant's. The red giant star Aldebaran has a surface temperature of 5000° above absolute zero and shines with 100 times the sun's true brightness. Aldebaran has a mass 5 or 10 times the sun's mass and a radius about 15 times the radius of the sun.

Red supergiants have cool surface layers that are tremendously spread out and rarefied. The red supergiant Betelgeuse, in Orion's shoulder, is so large that if it replaced the sun, the earth would be inside it! However, Betelgeuse's outer layers are far less dense than our own atmosphere. Betelgeuse has a surface temperature of 3500° absolute, a true brightness 2,000 times that of the sun, and a mass equal to about 10 solar masses.

White dwarfs are the small, dense endpoint of most stars' lives, barely the size of the earth and faintly glowing. We shall consider them in detail at the end of this chapter.

Some hot, blue, giant stars are surrounded by clouds of hydrogen gas. When photons from the hot star strike the hydrogen atoms around the star, each photon has enough energy to knock the electrons entirely away from the protons, thus ionizing the atoms in the gas. This would not be true for photons from the sun because most of **289** them do not have enough energy to ionize a hydrogen atom. Some stars eject part of their outer layers (mostly hydrogen) as they expand into the red giant phase. If the hydrogen gas lies relatively close to the hot star (that is, much closer than neighboring stars), we may see a "planetary nebula" of ionized gas surrounding the star.

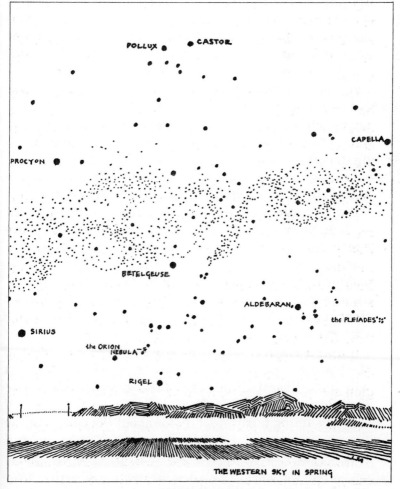

FIGURE 4

This term is a mistake, reflecting the fact that a planetary nebula looks round, like a planet. Such a nebula (this word is often used for any mass of diffuse gas) emits photons when the hydrogen atoms re-form themselves from separated protons and electrons, only to be destroyed again by new photons produced by the hot star at the center of the nebula. A typical planetary nebula like the Ring Nebula in the constellation Lyra has a diameter 10,000 times as large as the earth's orbit around the sun.[1]

If the gas lies relatively far from the hot star, that is, at distances comparable to the distances between stars, the energetic photons will produce a more diffuse region of ionized hydrogen which also glows by the light emitted when atoms re-form. The Orion Nebula, the middle "star" in Orion's sword (Fig. 3), actually consists of a region of hot hydrogen gas that is constantly being ionized by the photons from a group of hot, blue stars embedded within it. The hot ionized gas cloud has a diameter of about 50 light years. Such a hot cloud is called an H II region, reflecting an old terminology in which ordinary hydrogen atoms are H I and ionized hydrogen is H II. Stars like those in the Orion Nebula are extremely young (a few million years old at most), and apparently they are heating the rest of the gas cloud from which they contracted as proto-stars.

We can see that planetary nebulae and H II regions, despite their odd names, owe their light to the stars at their centers and thus to the same basic fusion processes that make the rest of the light in the sky. Now this can not go on forever. When the sun has processed most of the hydrogen nuclei in its central regions into helium, it will have liberated almost all the energy it ever will. Luckily for us, the supply of hydrogen in the sun will keep it shining fairly steadily for about the next 7 billion

---

[1]The distance from the sun to the next nearest star is 250,000 times the distance from the earth to the sun. One light year is equal to about 60,000 times the distance from the earth to the sun.

years, so the sun's total lifetime as a well-behaved star (adding what has already happened) should reach 12 billion years. More massive stars burn themselves out more quickly than this and less massive stars more slowly, as we described above.

Stars tend to contract their interiors as they fuse hydrogen nuclei (protons) into helium nuclei through strong and weak reactions that liberate energy of motion. The contraction is needed to allow the star to raise its interior temperature. The higher temperatures produce more rapid nuclear fusion because the higher velocities per particle allow the nuclei to overcome their mutual electromagnetic repulsion more easily. The more rapid nuclear processes allow the star to maintain its output of kinetic energy but at the price of consuming its nuclear fuel more quickly. After most of the hydrogen nuclei in a star's central regions have fused into He^4 nuclei, the star can contract still further, raise its interior temperatures, and fuse the He^4 nuclei into carbon 12 (C^{12}) nuclei.[1] Although we might expect that the star could then fuse C^{12} nuclei into still heavier nuclei, most stars do not manage to do this because at the point when most stars have fused He^4 nuclei into C^{12} nuclei, their interiors become degenerate.

The term *degenerate* refers to matter so *dense* that the rules of quantum mechanics make the matter *in bulk*, not just at the atomic level, behave in ways strange to our intuition. By the time that a star has fused most of its He^4 nuclei into C^{12} nuclei, the central part of the star will have contracted to a density thousands of times larger than its present density. This is the future for the sun about 7 billion years from now. When the densities rise to these great values, the electrons and the nuclei lie very close to one another. The motion of particles packed

---

[1]This process occurs in two stages: First, two He^4 nuclei fuse into a nucleus of beryllium 8 (Be^8), and then another He^4 nucleus fuses with the Be^8 nucleus to produce C^{12}.

close together is limited by the "exclusion principle" of quantum mechanics. This principle implies that *for certain kinds* of elementary particles and nuclei, no two *identical* particles can be in almost the same place *and* have almost[1] the same velocity. In stars, the elementary particles that are affected by the exclusion principle are electrons, protons, neutrons, and nuclei with *odd* atomic weights. As a star becomes denser, the first particles to be affected by the exclusion principle are the lightest particles, the electrons.[2] In a star 1 million times denser than our sun, the electrons tend to avoid one another, not only by electromagnetic repulsion, but also because the exclusion principle says they can not occupy almost the same point in space and have almost the same velocity. The exclusion principle in effect can produce a powerful mutual repulsion among electrons, far stronger than their mutual electromagnetic repulsion. This repulsion creates an additional outward pressure that helps the star to support itself against the contracting forces of gravity. A gas in which the exclusion principle plays an important role is said to be degenerate.

The exclusion principle will not become important for electrons unless the total density of the gas is at least 1 million times the density of water. In ordinary stars, such as our sun as it exists today, the exclusion principle does not affect the behavior of matter, but in highly contracted stars with extremely large densities the exclusion principle is an important effect. The exclusion principle keeps apart not only the electrons but also the positively charged nuclei. The electrons, kept immobile

---

[1] The "almost" in this theory has a precise, mathematical definition. In fact, the "quantum" in "quantum mechanics" is just this "almost."

[2] At a given temperature, each particle has about the same energy of motion, so the lightest particles (electrons) have the greatest velocities because the energy of motion is proportional to the mass of a particle times its velocity squared (see Chap. 6).

by their degenerate behavior, in turn keep the positively charged nuclei apart through electromagnetic forces. This occurs because each (immobile) negatively charged electron attracts and immobilizes the positively charged protons (Fig. 5), which restricts the motion of the positively charged nuclei. Because of this restriction the nuclei can not move fast enough to overcome *their* mutual repulsion and fuse, so nuclear fusion slowly stops in a star as its interior becomes degenerate. Within any star we would always find the positively charged nuclei

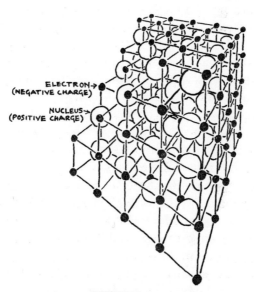

ELECTRON→
(NEGATIVE CHARGE)

NUCLEUS→
(POSITIVE CHARGE)

FIGURE 5

When matter reaches a density of 1 million times the density of water the exclusion principle affects the motion of electrons. No two electrons can be in almost the same place and have almost the same velocity. This effect of the exclusion principle tends to hold the electrons apart from each other. The electrons in turn hold the positively charged nuclei apart because of the electromagnetic forces between the electrons and the nuclei. Thus the degenerate behavior of the electrons provides support to both the electrons and the nuclei through the working of the exclusion principle. (Figure 1 on p. 322 shows the effect of the exclusion principle for the simpler case of a star made entirely of neutrons.)

and the negatively charged electrons thoroughly inter-mingled because any large-scale separation of positive from negative electric charges would produce an enor-mous attractive force seeking to reunite the two kinds of electric charge.

Mathematical calculations reveal that in stars with about the same mass as the sun the electrons will exhibit degenerate behavior after 10 to 20 billion years of hydrogen-to-helium fusion reactions, followed by less than 1 billion years of helium-to-carbon fusion. Stars with degenerate electrons in their interiors are not at all rare and form a class of stars called white dwarfs. A typical white dwarf, such as the tiny companion in orbit around the bright star Sirius, has a radius about the same as the earth's (one-hundredth of the sun's radius). Nonetheless, the *mass* of such a white dwarf almost equals the sun's mass, and a white dwarf is so dense that an average spoonful of the star would weigh 20 tons at the earth's surface! Inside the white dwarfs, nuclear fusion reactions stop almost entirely, and the stars cool slowly as the remaining energy of motion makes its way out through the dense material to the stars' surfaces, to be radiated away in the form of photons. This allows white dwarfs to shine faintly, with perhaps one-thousandth of the sun's true brightness, for billions of years.

For most stars, and probably for the sun, the end comes quietly and peacefully. Although it *may* have some minor flare-ups and instabilities, given enough time the sun should stop liberating energy entirely and will eventu-ally fade away, a degenerate cinder. We can not tell how many stars have become black dwarfs (not shining at all), but because it takes a long time to reach that state prob-ably only a few stars have dimmed out completely. Still, we can not be sure. A few percent of the stars that shine in our own galaxy are already white dwarfs, and there might be as many or more burned-out black dwarfs.

The term black dwarf usually refers to stars that have burned out quietly, ". . . not with a bang but a whimper,"

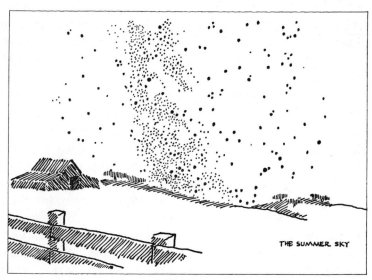

THE SUMMER SKY

FIGURE 6

as T. S. Eliot said. Since the burning-out process of nuclear fuel consumption takes longer and longer as the star grows dimmer and dimmer, the final stage of no light at all may not have been reached by any of the smaller stars (like our sun) that eventually become degenerate inside. But nature also shows examples of the opposite kind of star death: death by catastrophe. Such stellar blowouts are unusual, but they have a special importance, which we shall study in the next chapter.

SUMMARY

The sun and other stars shine by liberating energy of motion from energy of mass. In most stars, this energy liberation occurs through strong and weak nuclear reactions at the stars' centers. These reactions, called the proton-proton cycle, involve the most abundant nuclei in stellar interiors (protons and helium nuclei): the proton-proton cycle converts protons into helium nuclei while liberating energy of motion. This energy eventually reaches the surfaces of the stars, which radiate away energy of motion at the same rate as they liberate it in their centers. After billions of years of these fusion reactions, most of the protons in stars' centers will be fused into helium nuclei. Stars can then liberate a little more energy by fusing helium nuclei into carbon nuclei, but to produce the high temperatures needed for this fusion the stars' centers must grow very dense. After helium fuses into carbon, the density in most stars' centers has reached the point that matter there becomes degenerate.

Degeneracy restricts the motion of the lightest particles, the electrons, because the rules of quantum mechanics which apply to extreme densities state that identical particles like electrons can not have almost the same velocity and be in almost the same place. The electrons are effectively held apart by the rules of quantum mechanics that make dense matter degenerate. The electrons in turn hold the positively charged nuclei apart from each other by electromagnetic forces. This degeneracy pressure will keep nuclear reactions from occurring, and a degenerate star becomes a small, dense white dwarf, with about the size of the earth but of an average density 1 million times greater. White dwarf stars are rather common; they slowly radiate away their remaining store of energy of motion as they fade, on a time scale of billions of years.

QUESTIONS

1 What kinds of processes liberate energy in stars?
2 What kinds of forces are important in these processes?
3 How is the liberated energy transmitted through the star? How is the energy transmitted through space?

4 What kinds of elements (nuclei) are involved in the proton-proton cycle? Why do you think that other elements do not appear in the cycle?

5 What tends to hold a star together? What tends to make it expand?

6 If the insides of a star should liberate more energy than usual, why wouldn't the star explode?

7 Do photons produced in nuclear fusion reactions inside stars like the sun escape directly out of the star into space?

8 What happens to stars like the sun when the hydrogen nuclei (protons) in their centers have been transformed into helium nuclei?

9 What do we mean when we say that a star's central regions are degenerate? What causes this condition of degeneracy?

10 What effect does degeneracy have on a star's ability to perform nuclear fusion reactions? What does this mean for the star's future?

11 Why is the degeneracy of the electrons in a star important for the nuclei as well?

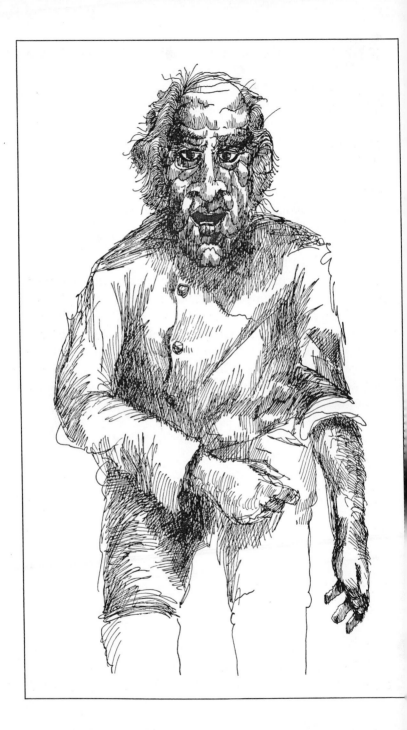

14
explosions
and
cosmic seeding

As the three Investigators entered the opening, the light from their beam illuminated a narrow staircase slanting sharply upwards. A cold and eerie breeze issued from the opening. Cyril Zaki's huge hands trembled slightly, but he flexed his pectorals, muttered "Inwards and Upwards," and let Card and Borg precede him up the constricting passageway. The first two dozen steps of the channel curved enough to block the last square of light from outside.

"What can you see, Julie?" Zenith asked anxiously.

"Nothing so far. Do we go on?"

"Push ahead, J. C.," said Cyril. "I've got the rear covered here."

The intrepid threesome hunched their way farther up the passage, penetrating deeper and deeper into the stone bowels of the structure.

"We must be almost to the other side," said Zaki. "I've counted more than two hundred steps."

"Congratulations," commented Zenith.

"Wait a minute," interrupted Julie Card. "There's something blocking the passage."

The staircase had widened just enough for Zaki, Card, and Borg to stand together and examine the obstacle before them. A huge block of stone that filled the channel *hung in front of them, seemingly suspended a foot from the stone floor. As the three investigators played their beam over it, the massive slab actually seemed to quiver.*

"Can you see under it, Julie?"

Card stooped down with her beam and had a look.

"There's an open space beyond there. Do we go in?"

"I think that's a job for our best man," Borg said. "Cyril, why don't you take the pocket torch and sidle under there?"

Why indeed, Cyril thought. How manly of me. Electing to go feet first, he moved crab fashion beneath ten tons of granite, sprang upright in the number four defense position, and peered craftily about him. Zaki saw that he stood in a surprisingly large chamber, thirty meters on a side, that rose to a vault ten meters above him. The air in the rock-hewn room held the musty, sealed-off smell of antiquity.

Cyril bent to speak under the stone slab.

"It looks all right," he reported. "But there's no other way out."

"Good risk," Zenith answered. "We're coming in."

In a moment the three were reunited inside the chamber. Zenith began to scan the walls carefully with her powerful beam. Strange pictures leapt from the darkness, long-armed men and women apparently reaching for a star.

"Aha," Cyril Zaki said. "I've seen that sort of thing before. On Rongbuk, for example, they say their star was about to explode and . . ."

"All right, Cyril, let's have your case histories some other time. What's over there?"

The lamp picked out a shiny black stone several meters square, almost completely covered with carved signs. Cyril walked over and had a close look.

"It's some sort of ancient writing, I think, and there's weird signs that don't look like letters at all."

Zenith Borg made a quick inspection and agreed.

"Right you are, Cyril. That's what they call mathematics. I can't understand the writings or the symbols though."

"But I can," said a calm voice from the dark.

"Zed!" hissed Zaki in a burst of rage.

"The archfiend himself," said Zenith, as she swung her beam toward the grey-haired doctor, who seemed remarkably sure of himself in his exposed position. "Julie, have you got your blaster aimed?"

"Right, Z. B. Say the word and he'll be junk."

Bokomoru Zed was unperturbed. "Have you noticed what I'm standing on here?" he asked his adversaries.

"Noooo," answered Captain Zaki. "It looks like a rock."

"Wrong again, Captain. It is actually a stone lever, and the ingenious designer of this chamber gave the person standing here the power to seal us all in forever. Now that I've removed the supporting stones, it is my weight alone that keeps the entrance slab suspended. So I suggest you sit down on those carved stone bathtubs over there and let me give you a bit of enlightenment. After all, Zenith, you've refused to listen for eighty years. Now that my life's work has ended in failure, you can afford to spend a moment hearing the facts. What I've found out during the past week will amaze even a closed-up power broker like yourself."

Borg decided to humor the doctor awhile and sat down on a conveniently located stone basin. Card and Zaki followed her example, and Zed smiled briefly before he spoke again.

"Do you know what the Permids really are, Zenith?"

"No, Bokomoru. Does it matter that much?"

Lost in deep reflection, Dr. Zed swayed from side to side on the stone lever, causing spasms of fear in Cyril Zaki. No sense being trapped inside here, Cyril thought. The worms would eat you for sure.

"Humanity began here on Thurd," Zed said. "The wall behind me is the master record of human progress, a kind of memory device. Every so often—I think every *ten thousand years—people would come here to chisel write on the slab over there to describe the events of that era." What a primitive accounting system, Cyril thought.*

No wonder they used it only rarely. Well, it was certainly mind bending to think of those oldtimers at work. Why was Zed so enthralled? The doctor continued:

"I've spent the last six days and nights deciphering the writing here, more than eight hundred entries. And not a single one of them mentions oofoes! I just can't fathom it—apparently humanity is all alone, alone, alone!"

30

Although most stars fade into an ever-dimming, degenerate obscurity, some stars end their energy-liberating lifetimes with a dying, cataclysmic, "supernova" explosion. Only about 1 star in 100 finishes its life with a supernova outburst, but when this happens it is worth watching. A single supernova within our galaxy can be bright enough to be seen even in the daytime for several months. The last such supernova appeared in the year 1604, and the supernova that produced the famous Crab Nebula (Fig. 1) was recorded by Chinese astronomers and native Americans in 1054 A.D.* An exploding supernova can liberate as much energy in 1 year as the star produced in all its previous millions or billions of years.

What determines whether or not a star will become an exploding supernova? The critical factor is the star's *mass*. If a star has a mass less than 1.2 times the mass of the sun, it will probably never explode. However, stars with masses greater than 1.2 solar masses can become supernovae. As we saw in the previous chapter, stars grow denser and denser in their interiors as they age. This inner contraction occurs because the diminishing supply of nuclear "fuel" demands a higher rate of nuclear reactions to maintain a steady liberation of kinetic energy. The steady rate of energy liberation is necessary to support the outer parts of the star against the forces of its self-gravitation that pull it together. To produce the greater rate of nuclear reactions the star must produce higher densities and temperatures by contracting its inner regions. That is, the star's inner regions gradually contract to avoid the catastrophic contraction of both its inner and outer regions which gravitational forces seek to produce. Calculations of the way which stars steadily contract their insides as they liberate kinetic energy show that only those stars whose mass is *less* than the critical mass of 1.2 times the sun's mass can exist in a

---

*European scientists, who were then suffering through the Middle Ages, failed to record anything noteworthy in 1054.

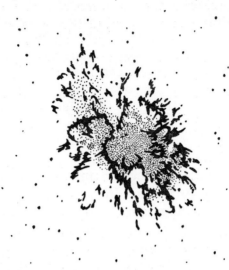

FIGURE 1

The Crab Nebula is the remnant of the supernova explosion seen in 1054 A.D. The nebula consists of hot gas and some extremely energetic particles, and there is also a strong magnetic field within it. The visible light radiation comes mostly from photons produced when particles move near the speed of light through strong magnetic fields. This kind of radiation is called synchrotron because it is also produced by particles accelerated in machines called synchrotrons.

stable degenerate state of matter (we discussed this degeneracy in the previous chapter). This critical value of the mass is large enough for most stars to fade peacefully as degenerate white dwarfs because most stars have masses equal to or less than the sun's mass.

Stars with masses much greater than the sun's tend never to become degenerate because more massive stars are *less dense* than less massive stars, and so a really massive star (say 5 or 10 times the sun's mass) never becomes dense enough to become degenerate. Matter becomes degenerate only when the density rises to a very large value. If two stars have the same temperature at their centers, then the more massive star will have a lower average density because the more massive star will have a *much* larger size than the less massive star. Most stars *do* have the same temperature in their central regions as long as they are liberating energy through the same series of nuclear reactions, and indeed most stars use the proton-proton cycle.[1] The more massive stars are less likely to become degenerate as they contract because the degeneracy is caused by the crowding together of particles, and the more massive stars are not as dense or crowded to begin with as less massive stars.

Let us consider the difference that degeneracy makes to a star as it grows older. Once a star has liberated energy of motion by fusing almost all its hydrogen nuclei (protons) into helium nuclei, the star can liberate some additional energy of motion by fusing helium nuclei into carbon nuclei. These more advanced nuclear reactions, involving nuclei with larger atomic numbers, transform a smaller percentage of the energy of mass into energy of motion.[2] In addition, these reactions require greater temperatures for the particles to overcome their mutual electromagnetic repulsion because nuclei with larger atomic numbers are more highly charged and repel each other more strongly.

Most stars reach the stage of fusing helium nuclei into carbon nuclei before the exclusion principle acts to keep the electrons apart and prevents the fusion reactions from occurring. To fuse helium nuclei into carbon nuclei the star's inner temperatures must be high enough (about

---

[1]Almost all stars now liberating energy of motion are doing so by hydrogen-to-helium fusion reactions. Most of these stars have central temperatures close to 13,000,000° absolute, even though their *surface* temperatures may vary from 2000 to 80,000° or more.

[2]The fusion of three He^4 nuclei into a C^{12} nucleus liberates one-fifteenth of 1 percent of the energy of mass. In contrast, the proton-proton cycle liberates about 1 percent of the energy of mass when four protons fuse into one He^4 nucleus.

200,000,000° absolute) for the helium nuclei to overcome their mutual electromagnetic repulsion. The star's insides reach this high temperature by contracting to produce large pressures and densities. In a star with a mass equal to or less than the sun's, the electrons will become degenerate when the densities and temperatures have grown large enough to make helium nuclei fuse into carbon nuclei. The next stage, the fusion of carbon nuclei into still heavier nuclei, requires temperatures of 600,000,000° absolute. Stars with masses less than 1.2 solar masses never reach this temperature and never fuse carbon nuclei into heavier nuclei because the exclusion principle stops the fusion reactions in the way described in the previous chapter.

Stars that have become degenerate in their interiors are in a fine position to do almost nothing. After nuclear fusion reactions cease inside a degenerate star, the star slowly radiates photons that gradually carry away the remaining energy of motion of the particles inside it. This slow leakage of energy can continue for billions of years because the rate at which the star radiates energy remains low. Our own sun, a typical star, will eventually become a degenerate white dwarf about the size of the earth, thousands of times dimmer than it is now. But what about the stars with masses greater than 1.2 times the sun's mass? Instead of ceasing all nuclear fusion reactions and using the exclusion principle for support against self-gravitation (as degenerate stars do), such massive stars increase the temperature in their interior and continue to fuse heavier and heavier nuclei. These stars can liberate some energy of motion by fusing carbon into heavier elements, forming oxygen, neon, magnesium, silicon, and iron. All these fusion reactions liberate some small energy of motion from the energy of mass, and this liberated energy presses outward against the star's self-gravitation. But when fusion reactions have built iron nuclei, the energy balance reverses. To fuse iron nuclei into still heavier nuclei, *additional* energy of motion must be supplied. For fusion reactions among iron nuclei, the products of the reactions have *more* mass than the original particles did.

For a massive star to support itself against the forces of gravity it must still liberate energy of motion in its interior. But how can it do so, when no more nuclear reactions that liberate energy of motion are available? The answer is that the star has a problem, and eventually the star's problem becomes overwhelming. We see that a star which wings its way through all the different steps of nuclear fusion is in for a rude shock. It can spend billions of years fusing hydrogen nuclei into helium nuclei, and millions more fusing most of its helium into carbon. From carbon the star can make oxygen and neon, and from these magnesium and aluminum, silicon, phosphorus, sulfur, and, finally, iron. At each stage the star's central regions grow hotter and hotter, denser and denser, to produce the fusion reactions by overcoming the mutual electromagnetic repulsion of the positively charged nuclei. All the while the forces of self-gravitation are causing the star to contract its interior more and more. Suddenly, the star finds that its insides are mostly iron, and there seems to be no way to liberate more energy of motion to oppose the contracting forces of gravity. At that moment, the central regions of the star *collapse*, in a time of about 1 sec, and this collapse in turn produces the supernova explosion.

The collapse of the star produces an extremely dense core, and matter falling onto this core from the star's outer layers bounces off and explodes into space at a high velocity.[1] This outward explosion off the dense core has a profound effect on the rest of the universe because it seeds the universe with heavier nuclei. The supernova *outburst* that follows the implosion of the star's center will spew "evolved" stellar material—carbon, oxygen, neon, silicon, iron, and so forth—into space. These heavy

---

[1]We discuss the remnant cores in the next chapter.

elements (by "heavy" we mean anything heavier than hydrogen or helium, both of which came from the original big bang) can then be included in a new generation of stars that later forms from the diffuse gas which includes the supernova-seeded heavy elements. Because the explosion process is so energetic, it can even produce some nuclei heavier than iron, such as gold, lead, mercury, and uranium.

All this works out just right because more massive stars complete their life cycles in a shorter time than do stars like the sun. A star's lifetime is roughly proportional to 1 over its mass squared (see page 288). Therefore, a star with 5 or 10 times the sun's mass should burn out itself much faster than the sun will. This could happen just 10 or 100 million years after the massive star first turned on. Such massive stars should spread much of their outer parts through space after they explode as supernovae. These stars should thus be responsible for the creation of all the elements heavier than helium, elements made in stellar interiors and seeded throughout space after the stars exploded because they never became degenerate.

When the solar system formed, the gas and dust from which it condensed apparently had been well sprinkled with elements heavier than helium. The sun is about 70 percent hydrogen (by mass) and 29 percent helium. About 1 percent of the sun's mass consists of nuclei heavier than helium. Some of the debris left when the sun condensed seems to have formed the planets and asteroids that orbit the sun as the "solar system." The most massive of these bodies, Jupiter, Saturn, Uranus, and Neptune, have gravitational forces large enough to hold on to most of their original hydrogen and helium. These planets are also much farther from the sun (5 to 30 times) than the earth, so they are colder than the earth. The lower temperature means that hydrogen and helium atoms on these planets will move more slowly than on earth, and thus they can

be held from escaping the planet more easily than atoms
on the earth's surface. The earth's gravity was too weak to hold much of the lightest elements (hydrogen and helium), so almost all the earth consists of elements heavier than helium, elements that we think were made in the last fiery gasp of a dying star.

COSMIC RAYS

Supernova explosions occur in other galaxies besides our own, and because there are many other galaxies we can hope to see some supernovae each year, even though in an average galaxy there is only one supernova about every 100 years. In fact, astronomers have observed many supernova outbursts in nearby galaxies. One such supernova was brighter (for a while) than all the other stars in its galaxy combined.

The enormous power of supernova explosions makes them the most likely source of the mysterious cosmic rays. These cosmic rays were named before people realized that the rays actually consist of ordinary nuclei—hydrogen, helium, carbon, oxygen, iron, and so forth, together with some electrons and positrons. In cosmic rays these nuclei, electrons, and positrons are moving with speeds very close to the speed of light, so their energy of motion is enormous.[1] A single cosmic-ray proton, for instance, which has a mass of about a trillionth of a trillionth (10^{-24}) of a gram,[2] can have an energy of motion of as much as 10 billion (10^{10}) ergs, equal to the energy of a well-hit golf ball. The huge energies of motion of individual particles in the cosmic rays made it hard for people to realize that they are actually familiar particles moving close to the speed of light.

---

[1] Chapter 6 shows how a particle's energy of motion is related to its mass and its velocity (see page 136).

[2] The total mass of the cosmic-ray particles is much less than the mass of the particles that are in stars in the universe.

Because cosmic-ray particles have such large energies, supernova explosions are considered a natural place to produce them since supernovae are the most impressive sources of quick energy bursts that we know. The outermost layers of an "evolved" star, which has become mostly iron at its center, should still contain some hydrogen, helium, carbon, and oxygen, as well as iron, so that a supernova explosion could accelerate some of these nuclei (as well as electrons, together with positrons and heavier-than-iron elements produced in the explosion process) to stupendous energies, thus making the cosmic rays that we detect here on earth.

In our own Milky Way galaxy, where on the average a supernova explosion occurs about every century, there should have been 100 million or so supernovae during the 10 billion years (more or less) that our galaxy has existed.[1] The particles in the outermost layers of exploding stars, spewn into space by supernovae in our own galaxy and in other galaxies, now form the separate component of the universe that we call cosmic rays. They owe their enormous energies of motion to the titanic power of exploding stars.

No one knows what produces the spontaneous variations (mutations) within a biological species that are the key to evolution through natural selection. Some scientists think that cosmic-ray particles may trigger changes within the genetic structure of cells that can be passed on to later generations. If cosmic rays govern evolutionary processes in this manner, then times of greater cosmic-ray flux, as when a supernova explodes relatively near the earth, would be times of more rapid biological evolution. The Russian astronomer Joseph Shklovsky

---

[1]Supernovae on the other side of our galaxy could remain unobserved (by us) because of the obscuring gas and dust, which is concentrated near the center of our galaxy. This might explain why no supernova has been seen in the Milky Way since 1604 A.D.

has suggested that the age of dinosaurs ended because of such a supernova explosion 100 million years ago; the explosion left the dinosaurs unable to deal with new factors in the struggle for survival.

SUMMARY

Supernova explosions appear in stars that fail to become degenerate. Most stars eventually become degenerate in their interiors and end their lives as white dwarfs a few thousand miles across and far dimmer than the sun. But if a star has a mass greater than 1.2 solar masses, it can not reach a stable degenerate state. Such stars continue to fuse nuclei in their interiors, not just helium and carbon (as less massive stars do before they become degenerate), but nuclei more complex than carbon. Inside these massive stars, nuclear fusion reactions proceed to make nuclei like oxygen, neon, magnesium, silicon, and so on, all the way to iron. Each fusion reaction liberates some energy of motion from energy of mass. However, to fuse iron nuclei requires *additional* energy of motion because the fusion products have *more* mass than the original iron nuclei. Thus a star with mostly iron nuclei in its central regions can not liberate more energy by fusion reactions. This situation induces a catastrophic collapse of the star's interior to form a superdense central core. The collapse breaks apart nuclei, fuses other new ones, and liberates a great burst of energy. The result is an explosion that blows the outer layers of the star into space. Such supernova events occur in our galaxy about every 100 years and produce stars bright enough to be seen in the daytime for a few months. Supernovae seed the rest of the galaxy with nuclei heavier than helium. They apparently also produce cosmic-ray particles, which consist of nuclei and electrons accelerated to enormous energies.

1 Do all stars explode at some point?
2 Is it likely that our sun will ever explode?
3 What kinds of stars become supernovae?
4 What is a star made of when it becomes degenerate?
5 Why do some stars not become degenerate?
6 If a star does not become degenerate, what does it do to liberate more energy of motion? Why does it need to perform this liberation of energy?
7 How does this affect the temperature and density in the star's interior? Why?
8 If a star does not become degenerate, what happens when the star runs out of ways to liberate energy of motion?
9 How big is a degenerate star? Is a supernova remnant larger or smaller?
10 Why are supernova explosions important to the rest of the universe?
11 Have the atoms in our bodies been made in stars? What parts of the atoms would you say were made in stars? How did they get to be found on earth?
12 What is unusual about cosmic-ray particles? What kinds of particles are they?

15
neutron stars and pulsars

A female voice from a dark corner interjected, "Boko, you zoron, haven't you figured that out yet?"

Zenith pointed her beam and revealed a small blond woman perfectly familiar to Cyril Zaki.

"My assistant," Zed explained. "No doubt overwrought by helping me with this mammoth translation effort."

"Assistant? Some kitzles you've got, to call me your assistant! Boko, I stuck around you to get some help with my research, you monomaniacal ego tripper. And aside from a bent for old languages, I'd say your mental power was smaller than your premium."

Cyril began to chuckle as he saw the hostility between his two former captors. Divide and conquer, he thought; too bad I didn't apply that fully back on Sidney. The assistant suddenly looked straight at Zaki.

"Cyril," she said, "it's time for you to know the truth: I'm Vibeke! I'm your sister, Cyril!"

"Vibeke? It can't be! She was taken by the worms!"

"That's just what ma told you, Cyril. Actually I was sent to Pilar for secret training after pa went to the funny farm. That's where I met Zed, and then I saw how Zenith Borg and the other 'Insurors' ran the Trust. But they caught me looking through the confidential policies file, and I was nearly sent to Solferino . . ."

"You!" ejaculated Borg. *"A troublemaker, a malconformer . . ."*

"But Boko here had worked out an escape plan, and we ended up on Sidney, working to rediscover the old knowledge. Of course, he had to think that he made the discoveries. Well, that doesn't matter now."

"Vibe, how can you say such things about me?" demanded Zed in anguish. *"Who was it who found out about the Mothers of Discovery? And . . ."*

"Don't waste time with that, Boko. Let me just make one thing perfectly clear. You've had oofoes in your brain for hundreds of years. Can't you see that any civilization more advanced than ours wouldn't waste time contacting us? Have you been trying to communicate with ants?

The doctor's finely tuned brain seemed to twang like a bowstring. *"You mean,"* he panted, *"they know we're here and they don't want to talk with us? Oh, awful! What junk this life is!"*

Cyril Zaki, already stunned to find his long-lost sister alive, was overwhelmed to realize that she had consorted with the evil Doctor Zed. What would ma think, he wondered, forgetting that ma hadn't thought much of anything for the past forty years.

"That's right, Boko," said Vibeke. *"I could have told you on Sidney, but I needed your assistance there because I wanted to find out the real history of the Trust. Why should oofoes mix themselves in our half-mad attempts to reach knowledge? Probably ninety-nine percent of all civilizations destroy themselves before they can get things straightened out. Every species for itself—that must be the glorious rule of the universe. Just look at the human record that we've pieced together here."*

With the two heretics within blast range, Zenith Borg had her fury well under control. *"Just what does the record tell you, then?"* she asked.

"Well, you probably know most of it, even if you'd made sure that people never learn the facts. Humanity doubled its numbers here, then doubled and doubled and

*doubled again. Once technology gave the chance to
extract the fossilized wealth of Thurd, people multiplied
like ants at a picnic. They gorged to the hilt on oil,
guano, coal, ripped out the copper and iron, aluminum
and sulfur. The race was on! By the time they noticed
they were destroying their future, it was far too late:
Grab all you can and never let up, someone's got less.
Getting to other planets became just part of the struggle,
war, waste, pollution, starvation—and it never stopped!"*

"Until Cyprian," Borg said cooly.

"Yes, Cyprian did something about it," Vibeke agreed.

*Cyril was puzzled. "Wasn't Cyprian the woman who
wrote all those videotape operas? How did that take care
of wars and starvation?"*

*Vibe looked at Zaki: cocky grin, widespread stance,
rippling muscles. Could he understand all this history?*

*"Well, there was a gigantic struggle for power every
ten generations or so. Each entrenched governing class
got thrown out eventually. Basically, the first five million
years of human history seem to be just the replacement
of one set of I-got-mines with another set, on a time scale
of a few thousand years. But finally Cyprian got her
chance. Have you heard of the Weald Wars?*

*"Sure," said Cyril, "like Billy Bumpass in 'Follow Me
Deadly.' I've seen that lots of times."*

*"Well, Cyril, most of the important history was left
out of that tape when the Trust made it. The leader of
humanity then was called Weinstein . . ."*

*"Oh, Weinstein!" Cyril interjected. "I've heard of him,
all right. You know: Weinstein/rama/Einstein/karma . . ."*

*"Beautiful, Cyril. Well, Weinstein insisted that the
struggle for power was essential—just natural processes
renewing the human survival instinct. Of course he had
to believe that; how else could he justify having gotten
into control? But Cyprian—now she had a real plan."*

"The insurance assurance," said Borg.

*"Precisely. Cyprian saw that only fear could compete
with greed in human beings. So after the Wars, she*

created a system to spread a little more ignorance and self-protectiveness through every generation, and she kept humanity from destroying itself. And that's not all, Cyril. It was Cyprian who discovered how to use the bahava in those melloroons you smoke!"

"Amazing!" Cyril said. "What a mind!"

"Well, it worked," Borg pointed out. "Fear and melloroons keep human rapacity in control to this day. Cyril here is an outstanding example of the way people are better off. He's the sort of man the Trust has always wanted to produce."

"Sure," Zed interjected. "He's the perfect specimen: a sound body in a sound body. Maybe you can live with that vision of humanity, but I can't. I say knowledge is far more important than the constipated contentment you've imposed for so long. Zenith, I'll give you a choice: either I drop the stone so you and I die here, or else we go back to Pilar and start writing a few new policies."

Zenith's grey eyes narrowed as she prepared a third alternative. Then she leapt to her feet, sprang onto the lever where Zed stood, and grabbed him by the throat. "You filthy degenerate heretic," she swore as Zed guggled helplessly, "Imagine a man like you trying extortion on me. I'll show you a policy, you uncalculated rotten doctor zoroo rat . . ."

But suddenly, with a move of fantastic suppleness for his aged body, Zed flung both himself and Zenith off the lever. Immediately a low-frequency rumble filled the chamber as the huge stone slab started to shiver its way toward the floor.

"Trust save us!" Zaki yelled, and he dove for the narrowing entranceway with such speed and grace that he was able to help pull Julie and Vibeke through in the last moments before the ancient mechanism of the sliding rock had sealed the chamber completely.

"Are you all right, Cyril?" asked Julie.

"Sure I am, but what will happen to Zenith? And Zed? We can't possibly get them out ourselves—and it

*will take years by Thurd time to get help from Pilar!
There's just no insurance here at all!"*

*"Well, Cyril," Vibeke pointed out, "They're inside
the burial chamber of the ancient rulers of Thurd. What
better place for the power-hungry of the universe?"*

*Julie and Cyril mused on this as they clambered
down the stone staircase. Zaki was especially impressed
by his memory that Borg's last words about him called
him the Trust's best hope. He reached the open air, sat
down on a massive stone block, and lit up a zinger to
watch the planet's illuminating star as it rose over the
wide river before him.*

"I've learned one thing," he commented after a moment.

"What's that, my brother?"

*"Melloroons will get you through times of no
insurance better than insurance will get you through
times of no melloroons."*

We think that a supernova explosion blows away the outer layers of a star when it goes off. The star's inner regions, which collapsed to start the explosion process, remain behind. They undergo a drastic contraction as part of the supernova process and become one of the strangest beasts in the universe: a neutron star.

Before a star explodes as a supernova, its central regions are not dense enough for the exclusion principle to be important; that is, the star is not degenerate inside.[1] The supernova occurs because the star has too much mass to reach a stable degenerate condition. To solve this problem, the supernova explosion blows away enough of the star's mass so that the star can become superdense and superdegenerate. A star must have less than 1.2 times the sun's mass before it can exist in a degenerate state. An ordinary white dwarf, the familiar example of a degenerate star, has a radius about equal to the earth's (4,000 miles) and a density about 1 million times the earth's density.[2] The collapse that follows as part of the supernova explosion process compresses the central regions of the star's remains into a radius of about only *6 miles.* Such a highly collapsed remnant will have a density of matter about 300 trillion (3×10^{14}) times that of the sun, or about 300 million (3×10^8) times the density in a white dwarf star. Table 1 gives the radius and average density of matter of the sun, a white dwarf, a neutron star, and a black hole; in this table the sun, white dwarf, neutron star, and black hole all have the same mass.

How is a neutron star formed? When an evolved star collapses to start a supernova explosion, the energy of motion in the collapse is so large that the colliding iron nuclei in the central regions break into a mixture of protons, helium nuclei, and neutrons. The energy of

---

[1]We discussed stellar degeneracy in Chap. 13 (page 291).

[2]The average density of matter in the earth is $5\frac{1}{2}$ times that of water; the sun has an average density almost equal to that of water.

TABLE 1 **321**

Different possibilities for a star with a mass equal to the sun's mass

Type of object	Radius, miles	Average density, gm/cm³
Sun	400,000	1
White dwarf	4,000	1,000,000 (10^6) (1 million)
Neutron star	6	300,000,000,000,000 (3×10^{14}) (300 trillion)
Black hole	2 or less	More than 8,000,000,000,000,000 (8×10^{15}) (8 million billion)

motion from the collapse provides the additional energy of mass needed to unfuse the iron nuclei into the protons, neutrons, and helium nuclei, which together have slightly more mass than the iron nuclei (we discussed this in the previous chapter). In the next stage of the collapse, the densities and pressures are so high that the helium nuclei break into protons and neutrons. During both of these stages of the collapse process there are enough electrons left inside the star to keep the total electric charge equal to zero. As the collapse proceeds still further, the remaining protons and electrons are squeezed together to form neutrons and neutrinos in the reaction

$$p + e^- \rightarrow n + v_e$$

At this point, the supernova remnant is so dense that the *neutrons* are degenerate. In fact, after the neutrinos escape from the collapsed remnant (since they have almost no interaction with matter), just about all that is left is a great number of neutrons squeezed into a few miles. The exclusion principle now operates on the degenerate neutrons and provides them with an outward pressure to resist gravitational collapse.

The entire star is now one giant crystal of regularly spaced neutrons, held apart by the exclusion principle (Fig. 1). Each neutron is restricted in its position and motion by the laws that govern degenerate matter. In free space, the neutrons would rapidly decay into protons, electrons, and antineutrinos, but in a neutron star these decays are immediately compensated for by the squeezing together of more protons and electrons to form more neutrons. Although most of the collapsed supernova remnant consists of neutrons, there are a few protons and electrons left in the mixture. The neutrons and protons can not fuse into heavier nuclei because the exclusion principle prevents fusion reactions from occurring. Thus the matter inside a neutron star will continue to be mostly neutrons for billions of years.

Two other properties of a neutron star are important: its rate of rotation and its magnetic field. Before the star collapses in a supernova explosion it has a certain amount

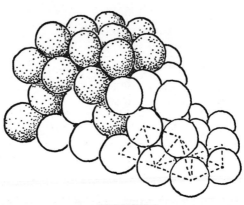

FIGURE 1
Inside a neutron star, the density is so large $(3 \times 10^{14} \text{ gm/cm}^3)$ that the exclusion principle strongly affects the motions of the neutrons which form the majority of the particles inside the star. The exclusion principle keeps the neutrons as far apart from one another as they can be because no two neutrons are allowed to be in almost the same place and have almost the same velocity. This degeneracy effect tends to hold the neutrons in a sort of lattice, with each neutron as far as possible from its neighbors.
The entire star is thus a sort of neutron crystal.

of spin momentum. As we discussed in Chap. 12, this spin momentum is proportional to the star's rate of spin times the square of the star's radius, and the spin momentum stays the same as the star collapses. Because the radius of the star shrinks by thousands of times as the star collapses, the star's rate of rotation must increase by *millions* of times to keep the spin momentum constant. Typically, a presupernova star may rotate once a month. A collapsed neutron star can rotate many times per second!

The magnetic field of the star also grows stronger during the collapse. The strength of the magnetic field at the star's surface increases as the star's radius decreases; this increase is proportional to 1 over the radius squared. A neutron star with a radius of 6 miles will have a magnetic field at its surface about 5 billion (5×10^9) times larger than the sun's magnetic field, if the star's original magnetic field had the same strength as the sun's.

The combination of the increases in the star's spin rate and in its magnetic field leads to some extraordinary results. A neutron star is a rapidly spinning, powerful magnet, and its spin and magnetic field are so large that their effects dominate the behavior of particles near the star's surface. Charged particles (electrons and protons) just outside the surface of a neutron star will be accelerated to enormous energies of motion by the electromagnetic forces generated by this giant spinning magnet. These particles in turn generate radio and light waves as they interact with the magnetic field. When fast-moving charged particles accelerate in a magnetic field, they produce photons (electromagnetic radiation) by the process called "synchrotron radiation" because photons produced this way were first observed in a type of elementary-particle accelerator called a "synchrotron." These synchrotron photons carry away some energy of motion from the rapidly moving charged particles that produce them. As a result, the charged particles gradually lose their energy of motion because they are producing electromagnetic radiation.

A neutron star produces photons of light and radio emission by accelerating particles in its spinning magnetic field. New particles are constantly being accelerated, only to radiate away their great energies of motion in the process called synchrotron radiation. Still more impressive is the fact that the radio and light emission from a neutron star *pulses* rhythmically as a result of the star's rotation. This constant rotation produces a "lighthouse effect." Because of the differences in the magnetic field strength from place to place near the surface of the neutron star, the synchrotron radiation varies in power from place to place. As the star spins around, the most intense regions of synchrotron emission give the effect of a rotating beacon. During the last few years, astronomers have discovered about 100 *pulsars*, so named because of the amazing regularity in the beat of the radio pulses by which they were first discovered. The best-studied pulsar lies at the center of the Crab Nebula (see Fig. 1, Chap. 14), and the pulsar is thought to be the remnant of the supernova explosion seen in 1054 A.D. This pulsar blinks on and off 30 times each second in both radio and visible light waves, which means that the radiation is coming from a neutron star rotating 30 times per second, if what we have said about neutron stars is indeed correct.

Neutron stars gradually lose some of their spin energy through the process of dragging their magnetic field through the charged particles around them. These charged particles (protons and electrons) are also remnants of original star that produced the supernova explosion. The magnetic field's constant rotation does accelerate some of these charged particles by electromagnetic forces between the magnetic field and the particles. This acceleration to high energies of motion produces the pulsating radio and light emission and sends some of the particles into space with huge energies of motion before they can radiate their energy away by the synchrotron emission process. Cosmic-ray protons with the greatest energies observed (millions or billions of ergs per particle) apparently come from pulsars, the collapsed remnants of supernova explosions. The lower-energy cosmic-ray particles may have been part of the original exploding layers of the outside of a supernova. That is, between the initial outward blast and the final collapsed remnant, supernovae seem to be responsible for all the particles in cosmic rays.

The process of accelerating charged particles robs the pulsar of some of its spin energy and causes the pulsar to spin a bit more slowly. Even though the Crab Nebula pulsar has been studied for only a few years, we have already found that it is slowing down by a tiny amount each year, so that in about 10,000 years it will be spinning only half as fast as it is now. All the other pulsars that have been discovered spin more slowly than the Crab Nebula pulsar, and these pulsars should all be the remnants of stars which exploded before 1054 A.D., perhaps before any written history existed on earth.[1]

SUMMARY

The remnant of a supernova explosion becomes a compressed object so small (6 miles in radius) and so dense (3×10^{14} gm/cm^3) that the particles in it are squeezed together to form neutrons. The dense packing of the neutrons makes them degenerate, so the exclusion principle holds the neutrons apart. In addition, the contraction of the supernova remnant from a radius thousands of times larger causes a great increase in the star's spin rate and in the magnetic field at its surface. The magnetic field of the rapidly spinning neutron star remnant

[1] No pulsar has been found in the part of the sky where the supernova of 1604 A.D. was seen. It appears likely that some supernova explosions do not leave behind neutron stars with magnetic fields strong enough to produce a detectable amount of radio and visible light waves.

will accelerate electrically charged particles near the star's surface by electromagnetic forces. This acceleration of charged particles produces synchrotron radiation as a series of regular pulses of light waves or radio waves. Thus the rotating neutron stars, the residues of supernova explosions, form the pulsars found in our galaxy. Also, the highest-energy cosmic ray particles apparently were accelerated out of the spinning magnetic fields of these pulsars.

QUESTIONS

1 What is a neutron star?
2 Why is it made mostly of neutrons?
3 Which is smaller, a neutron star or a black hole with the same mass?
4 What keeps a neutron star from collapsing under the influence of its own intense gravitational forces?
5 When an ordinary-sized star collapses into a neutron-star configuration, does it spin faster or slower after the star collapses?
6 Why does the spin rate change as the star collapses?
7 When pulsars pulse on and off with a definite frequency, what characteristic of the pulsar does this frequency reflect?
8 Do the processes that produce radiation from pulsars consist of nuclear fusion reactions? Do they involve electromagnetic forces? Do they involve weak reactions?

index

Donald Levy (left) and Donald Goldsmith met and began their professional collaboration at the University of California at Los Angeles during the 1963-1964 academic year. This book is one result.

about the book

This exciting and highly readable book presents the most interesting aspects of astronomy and modern physics in a way that can be understood by readers with little or no scientific orientation or motivation. Shunning the mass of confusing detail that characterizes the standard "textbook" approach, the volume combines an engaging science fiction story with clear, concise, and straightforward explanations of the important topics and exciting developments in physics, astrophysics, and astronomy. Designed for one-quarter or one-semester courses, FROM THE BLACK HOLE TO THE INFINITE UNIVERSE has been written with freshness and verve for today's students in secondary schools, colleges, and universities. There are no prerequisites, and more complex material is placed at the ends of chapters where it can be used and studied with maximum effectiveness by student and instructor. In addition, the informed and the would-be-informed layman will find in its pages an enhanced appreciation of the symmetry and mystery of the physical world in which we live.

Holden-Day, Inc. 500 Sansome Street, San Francisco, CA 94111

ISBN 0-8162-3323-3